# MANIPULATIVE MONKEYS

# Manipulative Monkeys
## The Capuchins of Lomas Barbudal

SUSAN PERRY

with JOSEPH H. MANSON

Harvard University Press

Cambridge, Massachusetts, and London, England

First Harvard University Press paperback edition, 2011

*Library of Congress Cataloging-in-Publication Data*

Perry, Susan, 1965–
  Manipulative monkeys : the capuchins of Lomas Barbudal / Susan
Perry, with Joseph H. Manson.
      p.   cm.
  Includes bibliographical references.
  ISBN 978-0-674-02664-3 (cloth: alk. paper)
  ISBN 978-0-674-06038-8 (pbk.)
  1. Capuchin monkeys—Behavior—Costa Rica—Reserva Biológica
Lomas Barbudal.   I. Manson, Joseph H.   II. Title.
  QL737.P925P47 2008
  599.8′5—dc22      2007020542

To Hannah Gilkenson, Julie Gros-Louis, and Wiebke Lammers, three *moneras supremas*, without whose aid this project would have been impossible

# Contents

# Prologue

When I am in the Costa Rican forest and have the opportunity to watch and listen to tourists or local farmers as they encounter monkeys in the forest, I am always struck by the profound difference in the way they perceive the animals, compared with my own perception. They seem to see the monkeys as clones of one another—as multiple copies of a particular species template that just happen to be near one another at the moment. Some are bigger than others, but other than that they are all alike. A few attribute emotions and intentions to the monkeys who make eye contact with them, but they are typically interpreted only as wanting to defecate on their observers or "throw" a stick at them. Most of the tourists and locals think that after five minutes of observing the monkeys, they have learned all there is to learn about these animals: they are black and white, furry, and live in trees. So far as I can tell from their comments, it never occurs to them that the monkeys are part of a complex network of alliances and social intrigue that extends through vast regions of forest, and that is influenced by decades of historical circumstances.

My own understanding of the complexity of capuchin social life has

2

been steadily built up over fifteen years of hard work, tracking changes in individuals' social circumstances and strategies as they have grown up, formed friendships, made enemies, and moved to new groups. Careful analysis of their patterns of social behavior has led me to understand not only that the monkeys interact in complex ways but also that they have an excellent understanding of the quality of other monkeys' social relationships. Now when I encounter a male monkey in the forest, I know a great deal about him: I know his parents, in most cases, as well as his brothers, sisters, cousins, and aunts. I also have hundreds of hours' worth of meticulous records of his social interactions and development. Most important, I know that he is not simply a resident of his present social group. His head is still filled with memories of monkeys he has known in other groups, and he also knows quite a bit about the lone males who will be challenging him for reproductive opportunities in his current group. There is no doubt about it: capuchin monkey social life is every bit as complicated as that of the Old World monkeys and apes. When I look at an individual monkey, I now see him or her not simply as an individual but as an active agent connected to hundreds of other individuals in an intricate and constantly shifting social structure.

Standing on a hilltop at my field site at Lomas Barbudal, looking for signs of monkey life hidden in the vast expanse of vegetation, I sense that the forest is alive with simian drama. I know that, somewhere out there, females are feverishly grooming the same female relatives with whom they have been allied for many years, while other group members tend to their infants. Meanwhile, their juvenile offspring are racing through the brush, wrestling with one another and perhaps inventing new games that will prepare them for the challenge of coalitionary politics once they are adults. And although they may seem at first glance to be less involved in the frenzy of social activity as they sit on the sidelines, the adult male capuchins are nervously eyeing allies for signs of treachery and keeping watch for opportunities to better their social position. As these males wander through the forest searching for potential allies or mating opportunities, they know that the reception they will receive from any particular individual can vary from warmest affiliation to lethal aggression, depending on who else is present when the encounter occurs. Thus their cognitive machinery is constantly whirring as they try to remember who is friends with whom and under what circumstances.

While I stand on the hilltop surveying the scenery below me, it is possible that two monkey groups are unknowingly drifting toward each other, destined to have an intergroup battle that will change capuchin history. It drives me crazy that I cannot know every fascinating detail of each monkey's life, but I am grateful that I have had the privilege of spending so many thousands of hours in their company.

I am also extraordinarily lucky to share this study with my husband, Joe Manson, who has been my closest collaborator throughout my research career. Few couples can work together as closely as we have, helping each other with data collection, coauthoring papers, giving lectures for each other in a pinch, commenting on each other's manuscripts, and of course helping each other with domestic tasks and child care. I could not have sustained such a long-term project if Joe had not provided me with companionship, morale boosts, an enthusiastic ear for monkey gossip, and, most of all, help raising our daughter. I am most grateful to Joe and our daughter, Kate, for giving up a more comfortable and stable life in the United States and Germany and accompanying me during my field seasons in Costa Rica, rather than forcing me to choose between career and family.

This book, like all of our work, is a joint effort. I wrote most of it, aside from Chapter 6 (on female-female relationships), which was written primarily by Joe, from Joe's perspective. So the word *I* generally means Susan, except in Chapter 6, where it refers to Joe. Like most married couples, we cannot resist interjecting our own comments in each other's stories. In particular, Joe also contributed to the discussions of alloparenting, mating systems, and lethal aggression.

What are capuchins, anyway? And why did I decide to study them? When Joe and I explain our work to people outside the small worlds of academic biology and anthropology, we usually cite a number of movies and TV shows that have featured capuchin monkeys: *Friends*, *Outbreak*, *Monkey Shines*, and *Monkey Trouble*, to name a few. Because of their cleverness and trainability, capuchins are popular as performers, as helpers for paraplegics, and as pets. Although they shine in these roles, they are not nearly as fascinating in captivity as they are in their native forests of Central and South America, doing the things for which natural selection has shaped them. Capuchin taxonomy is in a state of constant flux,[1] but most taxonomists recognize that the tufted, or brown, capuchins (e.g., *Cebus apella*, *C. libidinosus*, *C. nigritus*, *C. xanthosternos*) form one broad taxonomic group that is distinct from the nontufted capuchins (*C. capucinus*, *C. albifrons*, *C. olivaceus*, *C. kaapori*), and some taxonomists believe that these two groups represent different subgenera. Within each subgenus, there is virtually no geographic overlap between the species; however, most parts of South America are home to one species of tufted and one species of nontufted capuchin. I chose to study *Cebus capucinus*, the white-faced capuchin, not so much because it stood out in any way from the other species (at that time, very little was known about capuchin behavior in the wild), but because it resided in Costa Rica, a peace-loving country where we felt comfortable starting a long-term study. *Cebus capucinus* is the Central American capuchin, and it has a quite limited distribution that extends from Honduras down to the northern tip of Ecuador. There are reports of some range overlap with *C. albifrons* in Ecuador, but in most of its range *C. capucinus* is the only capuchin species. Most of what we know about capuchin monkeys comes from studies of *C. capucinus* and *C. apella*.

Capuchin monkeys have captivated the public with their curiosity, agility, and cleverness for hundreds of years. European travelers' reports from the fifteenth and sixteenth centuries comment on the curious behavior and appearance of these monkeys; the similarity between the coloring of the hair on their heads and the form of the cowls worn by capuchin monks is what gave the capuchin monkey its name. Reports of capuchins' tool use and elaborate foraging techniques date back 500 years, and the amazing ability of these animals to create and manipulate tools has been peppering obscure corners of the scientific literature ever since, though it has come to the forefront of media attention only in the past decade. Noted scientists such as Erasmus Darwin,[2] George Romanes,[3] Thomas Belt,[4] and Heinrich Klüver[5,6] were impressed by their manipulative ability, although an early prejudice against the intelligence of New World primates in comparison with Old World monkeys and apes kept these observations from being prominently featured.

In the 1980s and 1990s, primatologists such as ourselves began flocking to the neotropics in search of comparative data with which to test models of social evolution and the evolution of intelligence that had been developed out of empirical data on Old World monkeys and apes. Once we began systematically collecting information on this fascinating genus in the wild, there was no turning back: again and again, these monkeys challenged our preexisting notions about what monkeys should be doing. And we began to discover that capuchins had *independently* evolved many characteristics that are identical to those that biological anthropologists most want to explain in humans: they have enormous brains, and they are experts at making and using tools to get food; their varied and eclectic diets include not only fruit and insects but also meat, which is highly valued and shared. Capuchins combine a quick temper with an ability to perform quick mental calculations, such that they are formidable hunters and also will team up with allies to kill members of their own species. But their social skills are not all directed toward cruel Machiavellian plots: even the most aggressive males also cooperate in more benign contexts, such as the care of infants and the joint defense of the group against predators.

Because their social relationships are so important to them, capuchins, like humans, are constantly testing their bonds. As in humans, nonconceptive sex is often an important means of communicating about the quality of their relationships with one another. And capuchins are intellectually creative, with much of their creativity expressed through devising unique rituals for testing and maintaining the strength of their friendships. These rituals are transmitted within cliques of allies, with the result that traditions form. By the end of the 1990s, enough evidence for similarities between capuchins and humans had emerged that it was widely acknowledged that capuchins were one of the key taxa from which data were needed in order to understand critical aspects of human evolution. They had earned the title "the apes of the New World."

The Lomas Barbudal Monkey Project began in 1990 as a study of a single capuchin monkey troop. Since then (largely owing to the generosity of the Max Planck Institute) the project has expanded to include regular study of five monkey groups, and it involves a staff of up to ten or twelve researchers at any given time. Our research focus has been on social behavior, broadly defined, including topics such as the dynamics of social relationships, social cognition, communication, mating systems, social development, and social learning. Because we are anthropologists, we tend to gravitate toward those questions that have always fascinated evolutionary anthropologists trying to explain human origins.

Our colleagues Dorothy Fragaszy, Elisabetta Visalberghi, and Linda Fedigan recently published a landmark book, *The Complete Capuchin*,[1] which summarizes in exquisite detail all of the work that has been done on this remarkable genus prior to 2002. It is not our intention to replicate their work. Here, we focus almost entirely on the research done at our own study site, Lomas Barbudal, and we attempt to describe for the layperson what capuchin society is like, much as early ethnographers wrote vivid and enthralling descriptions of the human societies in which they worked. Although the work of early ethologists (researchers of animal behavior), such as Niko Tinbergen and Konrad Lorenz, was rich in natural history, the recent trend in scientific writing, even in animal behavior journals, has been to ruthlessly edit out anecdotes and descriptions of behavior and to rely strictly on quantification of behavior patterns to test particular hypotheses. In our experience, understanding any aspect of an animal's behavior is enriched by knowing as much as possible about the whole organism, as well as the society and physical environment in which it lives. In teaching primate behavior to undergraduates, we also find that students are more enthusiastic about the subject when we thoroughly immerse them in the details of just a few species, so that they can visualize what the animals' lives are like, rather than hopping from species to species every time we switch to a new theoretical topic. This book was written in part to help our students understand what it is like to be a capuchin monkey—and what it is like to do primatological research—but we hope that it will appeal to anyone who is interested in animals or in evolution. Most of all, we hope that it will be read by policymakers in the tropical countries where capuchin monkeys reside, so that they will appreciate the importance of these fascinating animals and assist in the drive to conserve their habitat.

Although we have attempted to limit our examples to just a few monkey families, to avoid overwhelming the reader with names, it was nonetheless necessary to introduce quite a few monkey characters to demonstrate the complexity of capuchin social life. To make the reader's life easier, we provide a "Cast of Characters" at the end of this book that gives information on each individual's position in the social organization. Also in the back of the book are a timeline of some of the key events in the history of two groups of Lomas Barbudal monkeys and a glossary of behavioral terms to guide the reader through the capuchin communicative repertoire. These additions will give the reader an idea of the number of individual identities and social relationships that each monkey must track over time in order to become a successful member of capuchin society. The first chapter describes in detail a typical day in the life of a monkey group, and subsequent chapters discuss particular social or foraging challenges that the monkeys must solve.

# CHAPTER 1

# All in a Day's Work

I jolt awake and look at my watch in a panic, thinking, as I do most nights, that I have overslept. It is 3:12 A.M. I could sleep another three minutes, but why bother? Silently, so as not to wake my husband, I pull on my green army pants, tucking them into woolen socks, and put on a long-sleeved shirt and a machete belt. I rush to the kitchen, ignoring the scurrying of roaches and mice as they scuttle away into various holes in the termite-eaten walls, and force myself to drink half a liter of warm milk. Ugh, what a way to start the day. I grab my smelly jungle boots, still wet and slimy from the previous day, pound them against the ground to evict any scorpions or ant colonies that have lodged there overnight, and put them on my feet. I buckle my leather snake leggings over those, pull on my fishing vest packed with gear, grab my backpack, and creep out the door, closing it as silently as I can.

I pull myself over the concrete wall that surrounds our house and push myself through the overgrown plants into the yard of the house next door, where I am greeted by the usual morning mayhem. Seven men and women are stumbling about in the dark, struggling with snake leggings, and muttering to one another. "Do you have a radio?" "Who has the

GPS?" "Does this cable work? Do you have the microphone?" Nick Parker is already in the driver's seat of the Land Cruiser, trying his best to look patient. Eva Wikberg is standing at the gate, ready to close it after we pull out. We pile into the back of the vehicle, and stack and restack the gear until it all fits. One person is still running in and out of the door repeatedly, in a panic, looking for something. I hear an exasperated sigh from someone in the car. "Come *on!* We're not going to get there in time!" hisses a voice in the dark. Finally the last body is squeezed into the car, and with great effort we manage to close the door.

For the next thirty minutes we sit tense and silent, crouched so as not to bash our heads on the ceiling as the car rattles over potholes in the dark. We are packed in like sardines, with backpacks weighing uncomfortably in our laps, keenly aware of one another's body odors. Someone in the front seat attempts to start a conversation and is greeted with surly silence. Nothing is funny at 4:00 A.M. The car passes by several pastures, drives by a few shacks and a couple of nicer cement houses, and then comes to a stone wall topped by a barbed wire fence. The car slows. "How are you going in?" asks Nick. "Past the tanks, or past the dogs?"

"*Donde los perros*" (past the dogs).

"Be careful—I'd leave the road before you get to the second fence. They probably won't follow you there."

"You've got a radio, right? We'll keep ours turned on. We should be able to hear you. *Suerte!*" (Good luck!) Mino Fuentes expertly climbs the fence, avoiding the barbs on the wire, and drops over the other side. In a few seconds, he has disappeared into the dark.

The Land Cruiser rattles on down the road, dropping other people off at various points along the way. It descends into a wooded valley and lurches to a stop by a river, pulling off into the trees just before the gravel road crosses through the water. As Nick and I climb out of the car, a troop of howler monkeys challenges our arrival, roaring at the top of their lungs and ending the display with a series of gurgles. Leaving the car behind, we jump a wooden fence and rush through the woods in the dark. "Where did you leave them again?" I ask after we have been walking at top speed for about twenty minutes.

"Near the three-hour-tour road. We'd better hurry."

We hurtle through the dark, brushing aside branches as we go and doing our best to keep an eye out for snake-like movements and patterns in the leaf litter. Nick curses and calls out, "Wasp nest!"

"Thanks!" I say, and I take a different route, which is less painful than his but plasters me in the face with a spider web. As we hear the sound of the waterfall up ahead, we slide down the riverbank into the cool water. I wade in to up my thighs, carefully negotiating the bottom of the river so as not to slip, while Nick tries the more daring approach of hopping from boulder to boulder in the dark, in a vain attempt to keep his feet dry at least until 6:00 A.M. I hear a small splash as he loses his footing and recovers his balance. "Ziplocs?" I ask as I double check to make sure my radio is sealed in its Ziploc bag before replacing it in my pocket.

"Yup," responds Nick.

I slosh up the far bank, water spewing out of the vents of my jungle boots. Glancing anxiously at the lightening sky, we rush through the forest for another fifteen minutes. Nick begins to stop and listen every fifty meters or so. "Think we should split up?" I ask.

"Yeah, maybe so. They should be here, but who knows—it was a full moon." The sound of crickets and frogs is punctuated by a scream. Both Nick and I jerk our heads in that direction. Another scream breaks out about fifty meters away. "I'll get that one," says Nick, as he crashes through the brush toward the second scream.

I pull out my Psion palmtop computer, set the time and date, and begin typing. "Abby's group. Location: 3-hour tour road/Cabuyo intersection. Ad lib: Scream, unknown individual." It's still too dark for binoculars to be much use, but I peer up into the canopy at the cat-size, curly-tailed black figure above me. The screamer and another monkey are silhouetted against the sky. They break into a dance, both of them grunting rhythmically and pacing back and forth, warily moving closer to each other. This continues for about a minute, and then the smaller of the two loses her cool and begins screaming again. Her dance partner lunges at her, and then two other monkeys run over to put in their two cents' worth. Now three monkeys are lunging at the screaming monkey, who is kneading her prehensile tail tip between her hands. "Why does the important stuff always happen before the sun comes up?" I joke to Nick, as he runs over.

Meanwhile, other pairs of monkeys start dancing, and a third fight breaks out. "Complete chaos, at 5:24:32 A.M.," I say in an exasperated tone into my microcassette recorder—we jokingly make a habit of recording the time to the precise second when these chaotic predawn events occur, since the time is the only aspect of the event that *can* be noted precisely.

"I think the victim is Opie," says Nick, squinting at the squirming, squealing monkey. "I'll keep an eye on her till it gets light."

"Watch out!" I cry, as a monkey breaks a branch above us. We both leap out of the way, and the branch falls between us.

"What's your problem, Thornhill?" Nick asks, as we gaze up at a familiar figure that is unmistakable even in silhouette, with his left ear missing and a snaggletooth protruding at a 45-degree angle through his right lip.

It is March 2003, and I am following a group of white-faced capuchin monkeys that I have studied for thirteen years. I came here to Lomas Barbudal, Costa Rica, as a graduate student in 1990, looking for an appropriate species in which to investigate the evolution of intelligence. Capuchins seemed ideal, since they have extraordinarily large brains (larger for their body size than any primate except humans) and have evolved this trait quite independently from apes. As I expected they would, capuchins turned out to be a fascinating study species, living in complex societies characterized by enough political intrigue to keep me fascinated for years on end.

Nick Parker is one of more than forty field assistants who have come to us from all over the world to volunteer for the project. Many leave after just a few weeks, horrified by the nuisances that plague monkey biologists in the tropics, but Nick has been with the project for nine months and is already an extremely skilled observer. This is a typical morning. We're already exhausted, even though our workday officially began just two minutes ago. And the monkeys, as usual, are cranky and belligerent. Fortunately, experience tells us that monkeys and observers will both get a second wind and be in an improved mood once the sun comes up and we've had some proper breakfast. I pull some cookies out of my pocket, and the monkeys start gnawing on sticks and guacimo fruits. I've only taken one bite, however, when the adult male Thornhill mounts Rain, a juvenile male, and the two of them begin grunting. Sex turns rapidly to play wrestling, and I abandon breakfast for a while to take notes on the computer.

By 6:20 A.M., the sun is high enough in the sky that we can reliably identify the monkeys. This year the main focus of the research is the social development of infants. The primary data collection protocol involves following a single monkey for ten minutes and recording all of its social behaviors, foraging behaviors, and self-directed behaviors. In addition, we record the comings and goings of all other monkeys within ten monkey body lengths of the focal animal, and also note behaviors of nearby monkeys that could affect the focal animal's behavior. Every two and a half minutes, and also at the beginning and end of the follow, my watch beeps to remind us to note the proximities of all individuals to the focal animal. Young monkeys move quickly at all times, and so it is really a two-person job. Both observers watch the monkey, so that the identification of other monkeys and the coding of the behaviors can be cross-checked on the spot. One person narrates aloud while the other types into the palmtop computer. If the action gets too quick for us to type and watch simultaneously, the data are recorded on a microcassette and then transcribed in the appropriate place in the file when the workday is over.

Our first focal animal for the day is Yasuni, the two-year-old son of Opie, the female who had been attacked first thing in the morning. I volunteer to be the spotter while Nick types. As we start the follow, Yasuni is alone and chewing on a stick, in the hope of finding insects inside. The stick is unoccupied, so he abandons it and pounces on a cluster of dry leaves. He slowly releases his grip on the leaves and his head darts forward as he catches and consumes an insect that was concealed inside the leaves. Then he scampers over to a young playmate who is foraging on guacimo fruits. He handles one, and then abandons the fragrant fruit uneaten and climbs into the adjacent *Bursera* tree, where he begins munching on that tree's hard, turpentine-flavored fruits. His gaze wanders to Toulouse and his nephew Rain, a pair of low-ranking juveniles. Yasuni glares at Toulouse, the younger of the two, and opens his mouth, baring his teeth in the stereotypical threat face.

Instead of defending his uncle, Toulouse, Rain runs over to Yasuni, puts his arm around his shoulders, and joins Yasuni in threatening him. Such coalitionary play is common in capuchins, and they have an elaborate repertoire of gestures designed to show who is siding with whom. Yasuni bounces menacingly at Toulouse, who seems completely unper-

Yasuni, a four-year old male, forages on *Cocoloba* fruits. Photo: Susan Perry.

turbed by the situation. Unable to get a rise out of their chosen victim, Yasuni and Rain part company. Yasuni bounds over to another pair of juvenile males and nuzzles one of them. Then he races over to his mother, pausing to see what she is doing, and runs back to Toulouse again, who has joined company with another infant male his own age. This time Yasuni is friendlier, approaching him and observing with intense but rather polite interest as Toulouse handles a leaf, looking for insects. Once an insect has been consumed, they drift apart again.

Yasuni wanders about, approaching and leaving various juvenile males who are foraging, and then comes over to stare me in the eye. Once again he walks over to Toulouse and his older adolescent brother Solo. Solo gives him a friendly cuff on the head, just as Nick's watch beeps to tell us that Yasuni's ten minutes are up. "Last point sample, activity social play, one length of Solo, Toulouse and Marañon, five lengths of Cassie and Bailey, and ten of Thornhill. End of follow," I say.

"That was pretty easy for a Yasuni follow," says Nick. Most follows of

Yasuni seem to involve high-speed chases of his best friend, Cassie, and wild play bouts in which five to ten other monkeys pile on top of him in a ball, wrestling and biting one another nonstop. After following these frenetic, play-crazed monkeys for just ten minutes, we feel like we have been collecting data for an hour, because so much happens in a few minutes' time.

When the focal follow is over, Nick and I split up, circulating through the group as we look for the next focal animal we need. As we wander through the group, we perform a "group scan," in which we mark the activity of each monkey we see and the spatial relationship of that monkey to every other monkey within ten body lengths of it. Each monkey is scanned no more than once per thirty-minute period. We use these data to get a sense of which monkeys most frequently associate with one another and also to determine how the type of activity affects proximity to particular individuals. For example, infants may prefer to sleep with their mothers, even though they spend much of their foraging time peering over the shoulders of adult males and most of their play time in the company of agemates who are more exciting than their mothers.

As we cycle through the group performing this scan, we also document anything interesting we see: for example, sex, aggression, grooming bouts, and interactions with other species of animals. At the moment, however, there is little action. Most of the group has moved into a big *Sloanea terniflora* tree to eat the fruits. *Sloanea* fruit is one of the most popular foods in the capuchin diet. The tree is, however, far less popular among the researchers and is currently the number-one most-hated field-site nuisance, defeating the snakes, wasps, and killer bees hands down. The big, buttressed trees are huge, with thick canopies that make visibility difficult. But worse yet, the fruits are coated with tiny purple hairs (which account for the Spanish common name of the tree: *terciopelo*, or velvet). To get at the tiny kiwi-flavored fruit inside, the monkeys must remove the hairs from the fruit's protective capsule and then bite off the end of the capsule. The hairs are virtually invisible to the human eye, and they are extremely sharp, like shards of fiberglass.

When the monkeys forage above us, the dislodged hairs float down in shimmering purple clouds and land on us, going straight through our clothes and lodging in our skin or, even worse, in our eyes. Once they are in your skin, you start itching like mad. The most frequently used attachment on a capuchinologist's Swiss Army knife is the tweezers, which we use about forty times a day to remove these annoying hairs. Nick and I had both been to the hospital in the past couple of months to have *Sloanea* hairs removed from our eyes, and so we reluctantly pull safety goggles from our packs and put them on, sweaty and annoying though it is to wear them while working in a tropical forest. One of the studies we are currently conducting is on the role of social learning in the acquisition of food-processing skills. Since *Sloanea* is one of the most difficult to process of the foods in the capuchin diet, it is a food for which we have developed a special data-collection protocol.

Nick sensibly edges over to the upwind side of the tree and calls out, "I'm starting with Al Gore." (Al Gore is a juvenile monkey, named by an enthusiastic Democrat. According to project tradition, whoever sees a newborn infant monkey first gets to name it.)

"OK," I respond. "How about I'll cover Vishnu, Jackson, and Diablita, and anyone else who comes up to the trunk of the tree, and you do everyone else?"

"Sounds good."

We pull out our microcassette recorders and get to work. I start with Jackson and begin to narrate into the recorder. "Jackson has plucked an entire twig full of fruits from the tree and holds it in his right hand, scrubbing the fruits against a branch. As he does this, his left hand swings back and forth in the opposite direction of the right hand, and it slaps the fruits every time the two hands pass each other. He is in ten body lengths of Vishnu, who is also processing, but neither of them is looking at the other." Jackson bites off the tip of a capsule and uses his fingers and tongue to pluck out the fruit and eat it. I pause and wait for Vishnu to select a new fruit and then record her movements: "Vishnu grasps a single fruit in the left hand and draws it down the branch in a long, slow stroke. Then she holds the fruit in both the left and right hands simultaneously, and pushes it up and down the branch repeatedly, using the force of her entire body to push it against the branch. Ten lengths of Jackson, neither looking at the other." Then I move on to Diablita, the alpha female. Like most adults, she has a set routine, efficient and inflexible. She selects an entire twig, holding onto the twig so as not to touch the annoying hairs on the fruit, and scrubs it quickly and rhythmically against a branch with her left hand. She does not look at anyone else in the tree but is focused on the foraging task.

Five lengths away, four-year-old Al Gore is still refining his technique. Like Diablita, he holds a twig in the left hand and scrubs. But midway through the task, he seems to get impatient with his progress and scrubs the hairy fruit between his two palms. He does manage to get some fruit via this technique, but he spends quite a bit of time wiping his palms on the branch to rid them of hairs that have lodged there. Undaunted, he picks another fruit and tries again. This time he starts by ineffectually pounding the fruit against a branch before resuming a scrubbing strategy. Now he is scrubbing as fast as he can with the right hand, and his left hand is spastically flailing about. It takes the juveniles several years to master this foraging technique, and some of their early attempts are hilarious. Sometimes, after watching their mom forage, infants will rub the hairy fruits all over their bodies, apparently understanding only that these fruits are interesting and should be scrubbed somewhere, but completely misunderstanding the purpose of the scrubbing.

After an hour or so, the last of the monkeys leaves the *Sloanea* tree, and we very gladly take off our goggles, brush off our clothes, and start picking hairs out of our skin. As I remove the teensy purple hairs, I also notice several virtually microscopic dots moving slowly up my shirt. "Uh-oh, Nick, better check for ticks. I seem to have walked through a nest."

"Can I have some masking tape?" he asks. I pull out a roll of wide masking tape, hand Nick a piece, and we both proceed to run tape all over our clothes. I go through several pieces of tape before I am satisfied that I have removed the dozens of little seed ticks from my pants and shirt. Nick has fared better, finding just six ticks.

The monkeys begin to socialize once more. Nut, a seven-year-old natal male (that is, a male still resident in the group in which he was born), grooms Thornhill. When Thornhill fails to reciprocate, Nut walks over to Tattle, a low-ranking adult female, and supplants her at her foraging spot; she screams and moves away from him. Thornhill stands up, arches his back, and fluffs out his hair. He places his hand between his legs and begins to urinate, splashing urine around in a boisterous way before strutting down the branch. Then he crouches and makes eye contact with Vandal, another adult female. She responds by pursing her lips so that they resemble a duck's bill and uttering a series of breathy squeaks that rise in pitch. Then she begins to grunt rhythmically. Thornhill also purses his lips in the "duck face" and pirouettes, never taking his eyes off Vandal. All of these vocalizations and gestures are part of the typical courtship display. It is hard to imagine a young, attractive female like Vandal being sexually attracted to someone as asymmetrical and decrepit looking as Thornhill, but of course the monkeys often surprise us. In this case, Vandal loses interest before a mating occurs. We move on to conduct more focal follows of infants.

The rest of the morning passes quickly as we move efficiently through our list of subjects, losing sight of only one focal animal before her ten minutes are complete. We hear an occasional gunshot, probably from someone hunting deer or peccaries, but it sounds like the poachers are too far away to worry about today. The monkeys meander slowly down the river, sometimes foraging in the canopy above the water, and sometimes moving into drier areas to forage on insects and guacimo fruits.

We pass right through a group of howler monkeys, who are napping, as usual. It is hard to believe that howlers and capuchins are from the same family—they could hardly be more different! Howlers spend about 80 percent of the daylight hours sleeping, and all of the night as well. When they are awake, they spend most of their time just eating, and only extraordinarily rarely do they engage in any social interactions. They just can't be bothered. Capuchins, in contrast, are rambunctious and feisty all day long and go out of their way to make trouble for anyone who crosses their path. "Go get 'em, Cassie!" chuckles Nick, as his favorite juvenile, a two-year-old male, charges toward a pair of snoozing howler monkeys and shakes the branch they are on, waking them up. They hoot a little bit and pull themselves up, backing down the branch away from Cassie. Thrilled with his success, Cassie once again lunges at the howlers, both of whom are twice his size. A subadult male capuchin, Till Eulenspiegel (named after the legendary German mischief-maker), rushes to Cassie's assistance and chases the hapless howlers out of the tree. Till forces an adult female howler to the end of the branch and glowers at her as she clings helplessly to a few tiny twigs and looks anxiously up toward him. Then he gives her a final push, and she falls a few meters to a branch below. Both Cassie and Till bounce and squeak at the howlers, wagging their tongues at them menacingly, as the howlers slink away.

Shortly after noon we come across a forest fire. A few days earlier a fire had begun on a nearby ranch and had burned quite a large section of forest before we discovered it. We had run to get assistance from Daniel Rojas, a *campesino* (farmer) who has been our staunchest ally in defending the monkeys' forest in the past few years. He brought rakes, shovels, buckets, and a *bomba*—a plastic "backpack" with a pump, designed for spraying water or insecticide on plants, and we spent the better part of the day cutting firebreaks with machetes and rakes, and hauling water from the river to put out the flames. Daniel, a short, wiry man in his sixties, puts us to shame in these situations by cutting *rondas* (firebreaks) three times faster than we can, and hauling far more water than we can manage. While we are sweating, dehydrating, and choking on the smoke, he cheerfully tells stories, makes good-natured jokes at our expense, and proceeds with the work without breaking a sweat.

Apparently we had not been entirely successful in putting out that fire, because now we have found a burning fallen tree that is threatening to spread the flames across the *ronda* to unburned portions of the forest. Sometimes the roots of burning trees are so hot that fires restart hours after they have apparently been doused. Setting aside data collection for a while, Nick and I chop a new firebreak around the crown of the tree with a machete and haul water rather ineffectually in Ziploc bags from the river—fortunately nearby—to extinguish the flames. The monkeys never react at all to flames and seem unfazed by the smoke.

An hour later, having accomplished all we can with just a machete and leaky Ziploc bags, we leave the still-steaming tree and resume data collection. The infants and juveniles are having a wild bout of play, scampering along the ground. As they rush past us, one of them reaches out a hand and playfully slaps Nick. Jackson, one of the males who had immigrated to Abby's group from nearby Chingo's group relatively recently, forages on guacimo fruit just a meter away from us. "His wound is healing pretty well. Did you see it?" asks Nick.

"No, not today," I answer.

"Hey, Jackson!" calls Nick, trying to get him to turn around so we can see his forehead. Nick makes all his funniest noises, trying in vain to get Jackson to turn around and look at us. But he is too busy eating to care about anything we do, and he knows from past experience that our vocalizations are not invitations to interact, or preludes to anything interesting. (On my project, we have a very strict policy of noninteraction with monkeys.)

His attitude toward us changes moments later, however, when the adult female named Vandal comes into view. Jackson suddenly fluffs up his hair, jerks his head toward Vandal, and bounces up and down on the ground in front of us, threatening Nick. Vandal starts threatening Nick as well and moves toward Jackson. We pull out the sound recording equipment and try to get some of the threat vocalizations on tape. Little Dali, the infant daughter of the alpha female, runs up to Jackson, squirming with excitement, and shakes her head back and forth rapidly, lips smacking and teeth chattering as she looks at him. Her immense clitoris is erect, and her hair stands on end as she produces a harsh, guttural series of gargling sounds. Inching closer to Jackson, she scrunches up her face like an accordion and gargles some more while staring at him. Dali's mother, Diablita, rushes over and threatens me. She is soon joined by her four-year-old niece Fishy, who also threatens me, pressing her cheek against her mother's as she does so. Fishy then climbs on top of her aunt, stacking her head on top of her aunt's like figures on a totem pole. Both threaten me again, and Diablita twitches and bounces at us. In her excitement, she grabs Fishy's toes and chews on them while glowering at us. Fishy's cousin Cookie rushes over to join the excitement and takes her place on Diablita's back, threatening us. Diablita makes strange, soft, growly sounds in her throat as she stares at us. Nick and I pretend not to notice their aggressive display, humorous though it is. Finally they get tired of threatening us and return to foraging.

The infants and juveniles keep us busy with their exuberant play during the next couple of hours of focal follows. This is normally a lot of fun, but today, instead of choosing a nice flat place to play, they are sliding up and down a cliff, which makes data collection difficult. Every now and then they all dash to the top of the cliff, and then they chase each other, sliding progressively lower and lower, until they make another mad dash to the top. There is really no good vantage point for us, especially since there is a tangle of virtually impenetrable vines midway up the cliff. We abort three focal follows in a row as the monkeys zip up and down the cliff with us clambering behind them, cursing and trying in vain to find some plant to hold on to that does not have thorns. The soil is dry and often gives way beneath us, causing minor avalanches. Finally

Nick stays at the top of the hill and I perch halfway down the cliffside, clinging to a sapling. I narrate the data into the recorder, with Nick radioing the action to me by walkie-talkie when the monkeys are out of my view for a few seconds.

While the youngsters play, adult female Maní stealthily courts Jackson. She follows him at a distance, rarely taking her eyes off him, and makes duck faces at him. Whenever he moves toward her, however, she loses her nerve and backs away a bit. A juvenile natal male, meanwhile, pursues Maní, making duck faces at her. The three of them form an odd procession.

Around 3:00 P.M., the monkeys once again troop into a grove of *Sloanea* trees for a second feast. This keeps them busy for the next several hours. Occasionally they make brief forays into shorter, leafless trees to look for insects. Cookie spends quite a bit of time searching for ants: she gently strokes sticks while watching for ants leaving their holes. When they run out, she bites open the sticks and quickly licks up the ants before they can bite her.

As the light dims, making it impossible for us to accurately record the details of the *Sloanea* processing, our conversation turns as always to the two burning issues of the evening: what will be served for dinner, and where will the monkeys sleep tonight? No doubt the other field assistants are also discussing these important matters (especially dinner) in various parts of the forest. Our breakfasts and lunches in the forest are rather dismal, since cooked food quickly rots in the tropical heat, putting us at risk of food poisoning if we eat lunch later than 9:00 A.M. Also, we cannot eat fresh fruit in the field without inciting feeding competition from the monkeys and putting them at risk of disease transmission from tourists who would want to feed them. So we mainly stick to peanut butter sandwiches, prepackaged cookies or crackers, and peanuts—snacks that the monkeys will not recognize as food, so that they will not view humans as potential food sources. This arrangement is good for the monkeys, but for us it is a terribly boring diet when we eat it every day for a year or more running. Hence the inordinate fascination with dinner, the one meal of the day that does not come out of a cellophane wrapper. "So who's home today?" I ask.

"Hmm, I think it's Matt and Tom. But who cooked last time?"

"Matt, remember? When I went to the *pulpería* yesterday to pay Norma for this month's groceries, she asked me why on earth Matt was buying thirty sweet peppers. I guess he bought every one she had in stock. Since locals use them only as condiments, not as a main dish, she was thoroughly mystified."

"That's funny. Is there anything Norma doesn't notice? I hope Tom makes lentil stew. Then again, it would be nice to have homemade tortillas, and those don't go with stew so well."

"I'm hoping for fried zucchini. It's group dinner tonight, so Joe will be bringing something too—I bet it'll be banana cake in the slow cooker."

"What are we talking about tonight?"

"Those papers on medicinal plant use in animals."

"Oh yeah, I read those last week. Have you ever read *The Shaman's Apprentice*? I'm reading that right now; it's really interesting."

"So do you think Cassie will be a great shaman someday?" I ask.

Nick laughs, "No doubt." Earlier in the year, the field assistants had been amusing themselves by arguing over the "nationalities" and "professions" of all the monkeys.

Cassie runs by on the ground, skids to a stop five feet from us, and pounces on Toulouse. Toulouse wrestles with him, somersaults away, and grabs a scorpion out of the leaf litter. He bites it, winces, bites it again, gets stung once more, and then chews it as fast as he can, contorting his face in a hilarious manner as the scorpion stings him over and over.

"No way would I take my cues from a capuchin shaman," I say, as I watch Toulouse suffering through his scorpion snack. "Look how painful their normal diet is! I hate to think how much worse their medicine would be. Have you ever tried biting the *Capsicum* peppers that they rub in their fur in the rainy season?"

"No, but Nando did one day when we were out together. The monkeys were so funny chewing them—their eyes were watering, and their noses were running."

"I only tried it once. I used up a whole water bottle trying to get my tongue to stop burning. I just don't see what is so enjoyable about the experience, but they go wild over those peppers."

"So where do you think they will sleep?"

"The *Sloanea* tree on Ridgeway stream."

"I bet they'll sleep on Daniel's Superhighway. They've been doing that a lot lately. Then they run up Pizote Ridge first thing in the morning. If they do that, we should have someone approach the top of the ridge from the fence line in the morning, and send someone else in on the trail below."

As the sun makes its rapid descent, the acacia trees close up their leaves and the birds' calls are replaced by the sounds of the night: frogs, crickets, and nightjars. The mosquitoes come out and begin to whine in our ears, settling all over us. I take off my hat and swat the backs of my legs, killing several with each swat. At least it is not the rainy season now. In the dry season, the mosquitoes tend to plague us only at dawn and dusk or near water, but during the other half of the year they are a constant nuisance, whining loudly into the microphone while we record, and stinging our hands so hard that we can hardly hold the binoculars steady.

The monkeys do a bit of frantic last-minute foraging before they file into the sleeping tree for the night. They choose a *Sloanea* on the Ridgeway stream origin, but near Daniel's Superhighway—we both made good guesses, but it will be a pain to find them in the morning. Satisfied that the monkeys are really bedding down for the night, we swiftly move toward the nearest trail and head for the car. As we depart, we hear some monkeys bickering over the best sleeping branches.

"Rambo Rambo Rambo?" Nick calls into his radio, as we approach the car.

"*Hola*, Skanky!" replies Mino.

"*Donde está*, Mino?" ("Where are you?")

"*Voy a salir por el camino, donde me dejó por la mañana. Los otros están por el* gravel pit!" ("I'm going to come out on the road, where you dropped me off this morning. The others are at the gravel pit.")

A female voice joins the conversation: "Hello! Rambo's, here!"

"Hey, Laura, where are you?"

"We're coming out of Water Source. Wait for us—we can meet you at the car."

"Oh, man, that is too close for comfort. I bet we'll have an intergroup encounter tomorrow. OK, see you soon."

We arrive at the car first and unlock it. Nick removes his snake leggings and tosses them into the back of the Land Cruiser. We peer into the dark. Laura Johnson and Eva come down the road. "What's new in Rambo's today?" I ask.

"Ohhhh, Aramis is in the dog house again."

"That's hardly news," I laugh.

"No, but it was so sad. He was chewing his tail all day. Every time he went in a fruit tree, they chased him out." We pile into the back of the Land Cruiser, and Nick starts the car. We rumble up the rocky hill.

"So how's Cassie doing?" asks Eva. "I haven't seen him in sooo long."

"Great!" says Nick. "He creamed some howlers today."

"He was having a bit of trouble with the *Sloanea* though," I said. "You would have thought he was practicing juggling instead of trying to eat it."

"Aww, come on! He's a genius!" laughs Nick. "Remember how he wrapped that spiny caterpillar in leaves before he rubbed it to get the spines off?" Cassie is Nick's favorite monkey, so he feels obliged to defend his honor.

"But what about the time he stuck his head in that plastic bag he found in the river and then started doing lost calls?" Eva chimes in.

"He wasn't being stupid, he was being creative!" Nick insists, grinning.

"Was it a lost call, or a scream he did? I keep hearing different versions of that story," I say. "Who saw it? Was it Hannah? Good thing we write this stuff down."

As we round a bend in the road, we are flagged down by Nando Campos and Juanca (Juan Carlos Ordoñez J.).

"How was your day?" I ask.

"Not too bad."

"Where are they sleeping?"

"By the killer bee boxes."

"Oh, that's not so bad. I was afraid you'd say the *quebrador* [gravel quarry]."

"How's Fonz?"

I can tell even in the dark that Nando is glowing with pride just thinking about his monkey hero. "Fonz rules. We had an intergroup with El Salto group. You know, the group with the really huge males. I hardly even saw them though, Fonz chased them away so fast. I got some great video of Fonz tooth grinding. It was *so loud!*" Nando has hardly gotten seated before he is rummaging in his backpack to find the video camera so he can show us Fonz's latest moment of glory.

The car slows again to pick up Mino. He makes a show of holding his nose as he climbs in the back with all of us sweaty, grimy people. "*Abre las ventanas!*" he says.

"Sorry, the windows are already open."

"Aaargh!!! *No way!* This can't be happening!" cries Laura as she doubles over.

"What?"

"I think some *Sloanea* just flew in the window and went in my eye. It's not fair! I'm not even in the forest right now! I just can't go back to the clinic. I've been in there for *Sloanea* removal three times this month."

"You have the worst luck. Maybe you should start sleeping in those goggles," I suggest. "Don't rub your eye. Maybe it'll come out on its own this time."

"Hey, Mino, did you have any luck?"

"*Por supuesto!*" Mino grins—Of course!—and he proudly pulls out a plastic vial and hands it to me. "*Caca de* Took!" We had sent Mino out on a very difficult mission this morning: his job was to track down a male who had migrated out of Abby's group and collect a fecal sample from him, so we could extract DNA from it and see if he had fathered any offspring while residing in Abby's group.

"Where did you find him?" I ask.

"*Cerca los tanques. El andaba con dos novias.*" ("Near the tanks. He was with two girlfriends.")

"Way to go, Took! I always liked him. Did you recognize either of the females he was with?" I asked. Mino wasn't positive, but he thought he had seen the two females with Chingo's group the previous year.

The ride home is always quite a different experience from the ride into the forest. Although we are still tired, and even dirtier and smellier than we were in the morning, we are bursting with monkey gossip that we want to share, in both Spanish and English to make sure that no interesting detail has been missed. Also, the ride is made more exciting by the fact that we must negotiate the Pan-American Highway for the last leg of the trip, and it is now full of lunatic drivers, all of them careening from one lane to the next to avoid potholes so that it is impossible to tell who is drunk and who is not. Distracted by conversation, we often hit an unexpected pothole, which sends the passengers' heads crashing into the roof.

As we reach the turnoff into town, the people in the backseat peer anxiously out the back window to determine whether the drivers behind us will interpret our left-turn signal as an invitation to pass us on the left. Indeed they do—someone screams out a "traffic alarm call," and Nick slams on the brakes to let a stream of cars pass us before he turns off the highway and onto our street. I jump out of the car and open the gate so that Nick can pull in. We pile out of the car and are greeted by the house menagerie, which consists of several dogs and crooked-tail cats that have been rescued from the street. Rulo, the one-eyed watchdog, wags his tail, and the puppies Tucker and Strider bound up to us, generously bearing putrid gifts of a rotting iguana carcass and a dirty diaper that they have found in their explorations around town. Everyone pauses in the hammocks to take off their snake leggings and boots, and to greet the animals. I go inside to fill out the data summary worksheet, saying what kinds and amounts of data were collected for each group that day. Tom Lord is waiting with the computer, ready to start dumping data from the palmtops into the laptop. Some of the field assistants come in and start doing exercises, as if thirteen hours of scrambling up and down cliffs wasn't enough exercise for one day.

I climb the wall again, heading to my own house for a shower. My four-year-old daughter, Kate, decked out in a pink ballerina tutu and a party hat, is bouncing and squealing at the door, holding her bedraggled Piglet doll. "Mommy, Mommy! Hurry and take a shower because I want to hug you, but I don't want to get monkey poop and bugs on me. Do you have lots of ticks on you today? I can bring you the tick tape. Piglet had a birthday party today. She is four. She got a new Quidditch broom for her birthday. See, she can fly! She's really good at it. And I helped Daddy make a banana cake! Yuck, your clothes are disgusting. Don't let them touch me." She stops bouncing for a second, looks remorseful, points at my muddy clothes, and adds, "Sorry, Mommy, I love you, but that is really gross."

I shower and remove the ticks that evaded me in the field, while Kate stands in the bathroom, chattering happily about her day. Then I dress and once again climb over the wall, this time with my family, Piglet, and a banana cake, to join the rest of the gang for our weekly group dinner.

Evenings are every bit as chaotic as mornings. Everyone dumps data into the laptop, passes equipment to the people who will be working in the field the next day, and discusses the logistics of who will be working with which monkey group. There is a lengthy conversation regarding the time at which we will have to wake up, and a strategy session about how to get everyone to their respective groups the next morning. Once that business is out of the way, we can proceed to discussing the "journal club" topic of the week and filling in the cooks and data cleaners on the monkey gossip of the day. There is no time to dawdle over dinner. We all still need to make lunches, fill water bottles, and pack gear for the following morning. I head home to read Kate some bedtime stories and to contend with the most critical of my e-mail messages, which Joe has kindly screened to save me time. Technology is really a mixed blessing. I think nostalgically about the early days of the project, when there was no phone and no electricity, and I was therefore exempt from all responsibilities other than following the monkeys.

# CHAPTER 2

# The Social Intelligence Debate and the Origins of the Lomas Barbudal Monkey Project

When I saw my first capuchin monkeys, a captive colony of brown capuchins (*Cebus apella*), I knew immediately that these were no ordinary monkeys. Though confined to a cage, they were constantly on the move, manipulating objects and boldly approaching the bars to examine me, turning their heads upside down to look at me from all angles. They radiated curiosity and a kind of focused energy. It was easy to imagine them telling each other that they had things to do—places to go and monkeys to see. I could hardly take my eyes off these assertive, busy animals, with their bulbous heads and punk hairdos.

The year was 1988 and I was at the Caribbean Primate Research Center (CPRC) facility in Sabana Seca, Puerto Rico. Joe and I, then recently engaged, were taking a break from our study of female mate choice in free-ranging rhesus macaques at the CPRC's Cayo Santiago facility, where Joe was conducting his doctoral research. Working on a well-known species at a long-established site had its advantages: for example, the Cayo monkeys' family relationships through female lines were known for as far back as thirty years. But I was itching to study a little-known species, so I could be certain of discovering things that no one else already knew.

Back in Ann Arbor, I began looking seriously for a thesis topic. The University of Michigan, where Joe and I were graduate students, was a major center for the study of the evolution of social behavior. Several faculty members and scores of graduate students were investigating social complexity and its psychological underpinnings in large-brained, group-living animals such as dolphins, crows, parrots, and apes—and, of course, humans—with the ultimate aim of understanding the evolution of human behavior. I became fascinated with the controversy then raging over the evolution of intelligence and its presumed anatomical basis, large brains, and I began searching the literature for little-known, large-brained primate species.[1] Given my very positive first impression of capuchin monkeys, I was pleased to discover that they had the largest brain, relative to their body size, of any nonhuman primate, with the possible exception of the closely related squirrel monkey.[2]

During the late 1980s, primatologists debated the merits of three hypotheses accounting for the evolution of large brains, a process that has occurred independently in several groups of mammals and birds. All three began with the observation that brain tissue must serve an important function because it is energetically very costly: for example, the human brain comprises 2 percent of a person's body weight yet consumes 20 percent of the body's energy. The first hypothesis was that frugivores (fruit eaters) need big brains to remember the locations of their food.[3] Fruits, unlike leaves and insects, are found in patches that are scattered widely over the landscape and are available for only a short time between ripening and disappearing—often into a competing frugivore's belly. The second hypothesis was that some primates need large brains to manage foraging tasks that involve complex eye-hand coordination and extraction tools, such as removing insects embedded in bark and removing fruit from hard shells.[4] The third hypothesis was that big brains enable primates to manage their complex social lives successfully.[5,6] It had become clear from many long-term studies in the wild that monkey and ape groups are remarkably intricate societies in which kinship, dominance, and "friendship" influence moment-to-moment interactions as well as enduring social patterns. As in human families and other social groups, no two pairs of individuals have exactly the same relationship.

At the time I was putting together my research plan, the frugivory hypothesis was the one best supported by comparative data. Among primates and bats, it was known that the species that ate more fruit had larger brains for their body size. Fruit is a particularly energy-rich food, so it could be argued that frugivory merely permitted brains to become larger over evolutionary time. But evolution does not produce traits just because they are possible; a trait must enable individuals to reproduce better than those lacking that trait. Primatologist Katie Milton had promoted an alternative "spatial memory" version of the frugivory hypothesis, based on her detailed comparative study of the behavior of large-brained frugivorous spider monkeys and small-brained folivorous (leaf-eating) howler monkeys in the same habitat, Barro Colorado Island in Panama.[3] At that site, ripe fruits are on the tree for only about a month out of the year for any particular species, whereas edible leaves are on the tree for at least six to seven months out of the year. So a folivore choosing a foraging path at random will have far greater foraging success than a frugivore following a random course. However, if a frugivore encounters ripe fruit and then makes a beeline for all the other trees of the same species, which are probably fruiting at the same time, then it can greatly enhance its foraging success. Frugivores require a much larger home range than do folivores, so remembering the identities and locations of each tree within their range can be a monumental task.

The second competing hypothesis for primate brain expansion, known as the extractive foraging hypothesis, grew out of anthropologists' fascination with elaborate tool manufacture and use as crucial aspects of human mental evolution. Sue Taylor Parker and Kathleen Gibson,[4] among others, had pointed to extractive foraging as promoting the evolution of enhanced sensorimotor coordination and hence large brain size. Many primates need to extract insect grubs from dead wood, or to open fruits that are protected by shells, spines, or irritating hairs. These primates have far greater manual dexterity than species that subsist on leaves or on fruit that can be simply plucked from the tree and stuffed into the mouth. Because most primates lack anatomical specializations, such as beaks, to help them forage extractively, the capacity to use tools is often present in primates who are omnivorous extractive foragers.

There was insufficient information to carry out a proper scientific test of the hypothesis that primate species that engage in more extractive foraging have larger brains than those that engage in less extractive foraging. But two genera (sets of closely related species) were known to have both conspicuously large brains and conspicuously impressive abilities for using tools to extract foods: chimpanzees and capuchins. Wild chimpanzees had long been famous for their tool-making and tool-using abilities, but capuchins had been seen to use tools only in captivity. However, it was known that wild capuchins were highly destructive foragers who tore apart trees while looking for insects.

The evolutionary hypothesis that intrigued me most was the social intelligence hypothesis, first proposed by Alison Jolly[5] in 1966 and later elaborated by Nick Humphreys[6] and a number of experimenters and fieldworkers, including Andrew Whiten, Richard Byrne, Dorothy Cheney, and Robert Seyfarth, to name a few.[7,8] They reasoned that individual primates living in complex social groups would have an evolutionary advantage if they could understand their own, and their companions', social relationships and use this understanding to advance their own interests. "Triadic awareness," a term coined by Frans de Waal to mean one animal's awareness of some quality of the social relationship between two other animals, captures the idea well.[9] Although Jolly is a lemur biologist, all the evidence strongly supporting the social intelligence hypothesis came from apes and Old World (African and Asian) monkeys, such as baboons, macaques, and vervets. For example, Frans de Waal described how three male chimpanzees used political skill rather than brute force to jockey for power in an intensely studied colony at the Arnhem Zoo in the Netherlands.[9] At times, the top-ranking male (the alpha male) needed the support of the third-ranking male to retain his status and had to "buy" his allegiance by ceding more mating opportunities to the third-ranked male than the alpha enjoyed himself. Coalitions were so important to the contenders that they sometimes attacked their own allies after they had remained neutral during a crucial fight, as if to punish them for their perfidy.

In one of many ingenious experiments on wild vervets, Seyfarth and Cheney played tapes of juveniles' distress calls from hidden speakers and found that the juveniles' mothers looked (not surprisingly) toward the speaker, but that other females looked toward the mother, as if they understood the relationship between the mother and her offspring. Studies of several species showed that grooming among female monkeys was directed up the dominance hierarchy more frequently than down, as if subordinates understood that it was to their advantage to curry favor with social superiors.

The concept of "theory of mind" was critical in the debates about the evolution of intelligence. Several researchers argued that an important breakthrough in cognitive evolution occurred when certain primates became able to recognize that other individuals did not necessarily share their own beliefs. Byrne and Whiten coined the term "Machiavellian intelligence" to describe their version of the social intelligence hypothesis, and emphasized the darker side of the phenomenon, what they called "tactical deception": the manipulation of others' beliefs to induce them to act in the interests of the deceiver. In a cross-species survey they found that, by far, the apes and baboons exhibited the highest rates of tactical deception.[10]

In one example of tactical deception, a young baboon was observing an adult female foraging on some grass corms, a nutritious resource he did not have the skill to obtain on his own. He screamed at the female, as if he was being attacked, and this brought his mother (the "social tool") running to his aid. The mother chased the other female away, leaving the youngster to feed on the unwillingly abandoned grass corms. Of course, incidents like this one invite heated debate over their interpretation. Some argue that they can be explained by simple processes in which theory of mind plays no part: this juvenile may once have screamed while really under attack during a corm-foraging session, discovered that his mother's approach incidentally drove competitors away, and so repeated the action later, in the same way that a laboratory rat learns to pull a lever to obtain food without understanding how the food-delivery apparatus works. But Byrne and Whiten argued that it was at least as plausible that apes and Old World monkeys were practicing true tactical deception, which in the corm-scam example entails understanding (a) what Mom will think when she hears the scream, (b) what effect her thoughts will have on her actions, and (c) how her actions will change the current situation (i.e., who controls the food) to the screamer's advantage.

Measuring social complexity has proven to be a daunting problem, and when I was designing my research plan no one had attempted to test systematically the hypothesis that primates with large brains have more complex societies than primates with smaller brains. Also, the quality of evidence for such slippery cognitive concepts as theory of mind was quite dubious for most species. Available information, however, led many primatologists to the conclusion that New World (South and Central American) monkeys had much less complex social lives than Old World monkeys or apes. Although much less was known about New World primates generally, they seemed to practice tactical deception rather rarely. Aggressive coalitions during within-group aggression were almost unheard of. Many New World monkeys were reported to have egalitarian relationships rather than dominance hierarchies. Robin Dunbar argued that, whereas apes and Old World monkeys exchange grooming for other favors, grooming among New World monkeys (and among prosimians such as lemurs) served only a hygienic function, not a social one.[11]

I was intrigued by and somewhat skeptical of the consignment of New World monkeys to the role of the social simpletons of the higher primates. Was it really the case that a book like Frans de Waal's *Chimpanzee Politics* could not be written about any New World primate? Most of our detailed information about New World monkey social behavior at that time came from just two genera, howler monkeys and muriquis. Perhaps these genera were unrepresentative. All New World monkeys are arboreal (tree-living), most are smaller and quicker than most Old World monkeys and apes, and many live in inhospitable habitats, such as swamp forest, so it was not surprising that detailed observations of their social behavior in the wild were scarce. And I suspected that much of the apparent difference between Old and New World monkey social behavior was the result of selection bias in the backgrounds and research topics of investigators of the two groups. Most researchers of New World primates were zoologists (indeed some had begun their careers studying bird behavior) and were interested in standard, nonanthropocentric topics in behavioral ecology, such as foraging strategies and mating systems. Anthropologists and psychologists interested in the evolution of humanlike traits such as social intelligence and politics gravitated toward the apes and those Old World monkeys that had already been shown to exhibit social complexity.

Having studied macaques and, before that, gorillas, I found it hard to imagine a group-living primate that did not practice complex social strategies. But perhaps capuchins would prove me wrong. They were already known to be frugivores and (at least in captivity) skilled extractive foragers, so presumably any data that I would collect on wild capuchins would simply support the frugivory and extractive-foraging hypotheses for brain expansion. However, if these large-brained monkeys failed to engage in social behavior at least as complex as that of baboons, macaques, and vervets, the social intelligence hypothesis would be challenged. I resolved to document the quality of social relationships in wild capuchins, and to seek evidence in these monkeys for triadic awareness and Machiavellian tendencies.

Once I had made up my mind to study capuchins, I started to look for an appropriate study site. Of all the countries where capuchin monkeys lived, Costa Rica seemed the most appealing: it was well developed, politically stable, friendly to researchers, and had an excellent public health system. I was fairly confident that I would want to establish a long-term field site, so all of these factors were important to me. I did not want my research interrupted by political unrest after I had invested a lot of effort in building up a demographic database. Also, I wanted a site where I could safely raise a family.

*Cebus capucinus*, the white-faced capuchin, was the only one of the four capuchin species that resided in Costa Rica. Colin Chapman, a primatologist who had conducted a census of the monkeys of northwestern Costa Rica, recommended Lomas Barbudal as a study site. Lomas Barbudal (commonly shortened to "Lomas") had recently been established as a biological reserve in the Costa Rican National Park System, largely through the efforts of Jutta and Gordon Frankie, the latter a University of California, Berkeley, entomologist known for his studies of bees.

Joe and I wanted to live together in the field and yet carve out separate research niches for ourselves, so we began devising a postdoctoral project for him to do in Costa Rica. The only other primate species at Lomas Barbudal besides the capuchin was the mantled howler monkey, and Joe began designing a study of feeding competition in howlers, in collaboration with Ken Glander, who had a long-term howler-monkey research site just down the road from Lomas Barbudal. Both Ken and the Frankies were extremely generous with their advice about working in Costa Rica. In early May 1990, shortly after we got married, Joe and I flew to Costa Rica to begin our pilot studies.

We still get many laughs when we recall our first months at Lomas.

Our graduate studies in anthropology at the University of Michigan had quite thoroughly prepared us to discuss the intricacies of evolutionary theory, and we knew plenty of trivia about linguistics, archaeology, and social customs in tribal cultures. But our training in biology was purely focused on theory and completely lacking in the practicalities necessary to set up a field site in the neotropics. We couldn't tell a bee from a wasp, identify a single plant family, or effectively use a compass or a machete. And we had been camping only once, on a quick trip to a state campground in Illinois (complete with a modern bathroom, of course).

When we arrived in Costa Rica, we arranged to rent a small house in Bagaces, the town nearest the reserve. With a population of about 2,000, Bagaces was a small, dusty town made up of three paved streets, a few general stores, some taverns with horses tied up outside them, and a bull-fighting ring. It was inhabited mainly by cowboys and farmhands who were employed by two large ranches nearby. Our house was a typical Costa Rican house: cement floor, wooden walls, and a tin roof, with a coconut tree and a cashew tree in the backyard. Fortunately, a Canadian fish biologist, Brian Wisenden, was in Bagaces finishing up his dissertation research during our first month of fieldwork, and he taught us some key Spanish phrases, such as, "Have you seen any monkeys?"

We had originally planned to commute from Bagaces to Lomas via motorcycle. But once we saw the condition of the roads near the reserve (eroded mud roads full of deep holes and rocks, and overgrown with shoulder-high grass), we realized that this was a foolish idea. For one thing, neither of us had ever ridden a motorcycle, and there was no good place to practice this skill. The Pan-American Highway, ten kilometers of which lay between Bagaces and Lomas, was chock-full of absolute lunatic drivers who would have thought nothing of running us right off the road. And the remainder of the route, six kilometers of gravel roads, was too much like an obstacle course to be navigated safely by beginners. We would probably have been reduced to roadkill before encountering our first monkey. Anxious to get out to the forest and at least catch a glimpse of the monkeys we wanted to study, we settled on a far less macho means of getting to work: a jeep-taxi.

People often ask me how I chose Abby's group (the group of capuchins that was ultimately to become my main study group). It all came about via serendipity rather than careful planning. Joe and I got in a taxi and asked the driver to take us to Lomas Barbudal, preferably to a place where he had seen monkeys. He drove us out into the forest and deposited us at the side of the road, right before the road met a river. We made arrangements for him to pick us up at the end of the day. As we were to discover many months later, this place was not, in fact, part of Lomas Barbudal Biological Reserve: the river, Río Cabuyo, is the reserve's northern boundary. The taxi driver very sensibly did not want to damage his car by fording the river and negotiating the much worse roads in the reserve itself, so he left us there, figuring (correctly) that we would be none the wiser. We were so eager to begin our research that we did not shop around for more convenient sites but set about familiarizing ourselves with the first monkeys we saw.

So—there we were, in our first tropical forest. We had no map. We had no idea which plants were poisonous, which insects were dangerous, or what creatures were making all the sounds we could hear. We proceeded cautiously into the forest, mostly staying on the one trail we had been shown, keeping the Río Cabuyo in view so we wouldn't get lost, and avoiding touching plants that looked suspicious. The forest did not look at all like the forests we knew. There were virtually no tall straight trees—all the trunks were gnarled and twisted into odd shapes. Except near the river, the trees were fairly short (less than twenty meters high), and the crowns were thick with vine tangles connecting one tree to another. Away from the trail, it was difficult to walk in a straight line because the ground was thick with herbaceous plants, most of which were covered in thorns, burs, or nasty hairs.

One of the most jarring aspects of the vegetation was the prevalence of terrestrial bromeliads and various forms of cactus, which looked more like something we would expect in a desert than in a tropical forest. Lomas Barbudal is one of the few remaining Pacific dry forests. Dry forests are far more endangered than rain forests, and their appearance, aside from the cactus and the prevalent vines, is actually more similar to the deciduous forests familiar to North Americans than to a true rain forest. A few animals well-known in North America, such as the white-tailed deer, the raccoon, the skunk, and the possum, inhabit the Pacific dry forest alongside the monkeys, ocelots, kinkajous, and other "jungle animals." (Oddly, at least from our perspective, the white-tailed deer are considered by Costa Rican locals to be far more exotic and beautiful than the monkeys, which are taken for granted.)

Once we were twenty meters or so from the river, the vegetation changed from broad-leaf evergreens to deciduous trees, which were leafless because it was the end of the dry season. In the absence of a leafy canopy, the sun beat down on us unmercifully, sending us scurrying for the river and relief from the oppressive heat after just a few minutes of walking in the open. Nevertheless, we could readily imagine that observation conditions would be wonderful—if we could just find some monkeys. But we had no idea where to look for them, and even if we had known where they were, the ground was covered with leaf litter that crunched as we walked, making it impossible to sneak up on any animals.

We had been walking for only a couple of hours when we saw our first family of howler monkeys, a nice-size group of thirteen animals. Knowing very little about their behavior, we had expected them to run from us. So we were naïvely delighted by the "trust" (in fact, the laziness) these animals exhibited when seeing us for the first time. We stood entranced for an hour or two, examining every detail of their bodies for marks that would help us recognize them as individuals in the future. There were not many useful markings—they were all black-faced and scar-free, though a few had faint orange discolorations on the soles of their feet and hands. The monkeys gazed at us in a vaguely worried way for a few minutes, howled a bit, and then resumed their naps. After a while, we tore ourselves away to look for some capuchins. Continuing down the river for two more hours, we found no sign of the capuchins and finally came back to watch the howlers some more.

Day after day passed in the same way. Joe invariably found one or more groups of howlers, and I found no capuchins. I occupied my time by trying to learn to identify the plants and animals I found, using Dan Janzen's *Costa Rican Natural History* as a guide.[12] With Joe's help, I also began crudely mapping the course of the Cabuyo and the small streams feeding into it, so that I could orient myself better. Little by little, I gained confidence and began making longer forays away from the river without getting too terribly lost. I also began to help Joe mark the trees that he was going to identify, measure, and monitor as part of his study of howler monkey feeding strategies. This was often more exciting than we intended, as our hammering against a tree to insert a tag frequently disturbed the insects living inside the tree. Since we had no idea what kinds of insects these were, we got many an adrenaline buzz as we fled in haste from a partially tagged tree, pursued by swarms of aggressive social insects, often abandoning a hammer or backpack in our hurry to get away without too many stings. Gradually we learned through trial and error which insects were horrifyingly nasty (like the Africanized honeybees—i.e., "killer bees"—and certain wasp species that would chase you for up to a kilometer) and which were benign (like the termites, which had impressively large nests but were completely unaggressive unless you made direct contact with them). The term "beehive hairdo" acquired a new meaning for us, now that we regularly had to pull dozens of angrily buzzing sweat bees out of our hair while sprinting away from their nests.

After a week or so of searching in vain for capuchins, I began to entertain fantasies of working with howlers or the far more interesting raccoonlike coatis that I often saw. Finally, however, I saw my first capuchins. Hearing some unfamiliar cawing noises (which I now know were alarm calls directed at me), I looked out over a wide bromeliad patch and saw some tiny creatures bouncing up and down in the branches, glowering at me. I hastily put my binoculars to my face and confirmed that they were capuchins—at last! It was love at first sight for me, though certainly not for them.

They could not possibly have been more different from the phlegmatic howlers. Every group member was alert and in constant motion. The males in particular were vehement in their simultaneous disapproval of and curiosity about me. They turned their heads this way and that, to get a look at me from all angles. They stacked on top of each other so that their heads resembled totem poles, demonstrating that they were forming a united front against me. They uttered squeaky little threat vocalizations that sounded anything but menacing. Their capacity to break large branches was truly impressive, however. I began to wish that my mentors at Michigan had recommended a hard hat as an essential piece of field gear, because I could not imagine having the nerve to walk directly under a group of monkeys that was behaving in this way. Even in my first brief encounter with the monkeys, I could distinguish some individuals from the others. Many had scars, and there was quite a bit of variation in the shape of their black "caps." After about thirty minutes of displaying at me, the capuchins sped away, glowering back at me over their shoulders. I tried in vain to cross the sea of spiny bromeliads between us but could not manage it in time to keep up with them. Nonetheless, I was exhilarated by meeting the capuchins and confident that they would be interesting enough to justify whatever difficulties they would present me with.

My next few encounters with them were also brief, but each one lasted a bit longer. I soon realized that I would have to stop being so cautious if I wanted to keep the monkeys in view. So I abandoned my policy of assuming that every plant in my path was potentially toxic (luckily, as it turns out, there are no poison ivy–like plants in this forest) and scrambled up each hill the monkeys climbed, no matter how many wasp stings I received. Likewise, I waded through all the rivers and thrashed my way through the thorn bushes as best I could. At first I was able to break my way through most of the brittle vegetation, but after the rains began in late May, the undergrowth flourished and the vines became stronger and more resilient. It was clear that I was going to need a machete if I was going to keep up with the monkeys, and I would also need to create some more trails.

With great trepidation, Joe and I tried to buy a machete. There was no equivalent to a K-Mart or a supermarket in Bagaces. There were only the *pulperías*, like old-fashioned general stores, in which all the goods are kept behind the counter. If you didn't know the word for the product you wanted, you were out of luck. During our very first trip to the *pulpería* nearest our house, we were alarmed to discover that our previous experience in Latin America was much less valuable than we'd hoped. When we asked for *habichuelas*, the word for "beans" in Puerto Rico, we were greeted with a look of suspicion and bafflement from the store owner, who apparently had never met a foreigner. After a few minutes of trying unsuccessfully to communicate, meanwhile contemplating a protein-free diet, I spied a bag of beans behind the counter and pointed frantically at it, whereupon I was informed that these were *frijoles*.

We were relieved to find that a *machete* is a machete, even in Costa Rican Spanish. As lifelong city dwellers, we thought a machete looked more like a lethal weapon than a garden tool, and we felt extremely self-conscious carrying it home, half expecting to be stopped by the police. But of course carrying a machete is a normal thing to do in Bagaces. We also did not realize until we tried it out in the forest that a machete needs to be sharpened first, and it took many more rounds with the general store to sort out the correct terminology for "machete sharpener." Finally we persuaded an amused *campesino* to show us how to use a *lima* (file) to sharpen a machete, and then we were finally in business. We were nowhere near as efficient at trail cutting as the locals were, yet slowly but surely we began to expand the trail system beyond the one riverside trail that was maintained by the local community.

Even though I was getting better at following the monkeys, the problem still remained that I lost them each night when we left the forest. Staying with them until dark, when they bedded down for the night, did no good, since they were always long gone by the time I arrived in the taxi the next morning. (The taxi driver soundly rejected our proposal that he drive us to work at 4:00 A.M. each day.) Reluctantly we came to the conclusion that we would have to camp in the forest. We began arranging for the taxi driver to drop us off for a few nights, so that I could run out to the monkeys before dawn and thereby avoid losing them every day. Once we lost them, it often took several days to relocate them, and I could not afford to lose so much work time. We had no cooking equipment, and we got very tired of canned bean paste and tuna that summer. But it was wonderful being out in the forest at night, watching the bats foraging in the moonlight and being awakened by a chorus of howler monkeys just before dawn.

By the end of the second month I was beginning to get a feel for the social dynamics of the capuchin group and the personalities of the individual monkeys. They still would not come very close to me, and they always kept an eye on me. So although I was collecting data, it was just for practice; I could not use it for analysis yet, as they were clearly not exhibiting their normal behavior. I was already caught up in the monkey drama, though. The leader of the group was a mottled old male whom I dubbed Curmudgeon because of the perpetual snarl on his face and his incessant scolding of me. (He also bore a strong physical resemblance to a cantankerous professor I had once known.) He was the obvious favorite of the females, who frequently groomed him. Wherever Curmudgeon went, he was followed by a crowd of excited fans, most of them infants or young juveniles, who made a horrible gargling sound as they moved in to try to touch their idol. Although this gargling vocalization reminded me of a rhesus monkey threat, capuchins seemed to intend it as a friendly signal. Everything Curmudgeon did, he performed with flair and loud vocal commentary. For instance, every time he urinated, he arched his back, fluffed up his hair, made a resonant rhythmic grunting sound, and splashed urine all over his hands and the monkeys in the vicinity. None of the other monkeys looked anywhere near as important when they urinated.

There were two other adult males in the group as well. Paul Bunyan was a large male, with a rather vacant expression on his face and a drooping jaw, who had few social contacts and spent most of his time engaged in vigorous deforestation. He simply couldn't pass up an opportunity to strip bark off a tree or break a branch. Guapo, the third male, was a gorgeous, youthful-looking male without a scar on his body. He had an exceptionally fine hairdo—a perfectly symmetrical cap with a fine, neat point on it. Guapo was highly social and spent inordinate amounts of time playing all sorts of crazy, creative games with the juveniles. Sometimes he would become too exuberant and lose his self-control, causing a playmate to scream for help, but most of the time he was enormously popular with everyone but the high-ranking females, all of whom preferred Curmudgeon.

The oldest-looking member of the group was an adult female, whom I named Abby (short for *abuela*, grandma). She was a large monkey with big fluffy brows and a dark face. She could always be found in the center of the group with her sidekick, Squint, a feisty monkey who closely resembled Abby except for a scar on the side of her head that made one eye squint in a piratical manner. Abby and Squint were virtually inseparable and could absolutely count on each other for aid in any context. Abby was obviously of a higher rank than Squint, but she had a placid temperament and was not terribly prone to meddling in others' affairs. Squint, on the other hand, was so busy monitoring everyone else's social interactions that I sometimes wondered whether she spent enough time eating. She was easily outraged and had an opinion about everything. If an adult male so much as looked at her the wrong way, she screamed at him and solicited Abby's aid. Together they made a formidable team, and they could evict anyone from the feeding trees. Later, after analyzing my data, I was able to partially confirm my suspicion that Squint was the most socially astute member of the group. When I created matrices of social favors exchanged, it turned out that Squint distributed her social favors, such as grooming and coalitionary aid, precisely in the manner one would predict if she were calculating favors received and returning them in accordance with that knowledge. She also took rank into account, such that she both gave and received social favors with higher-ranking monkeys more than with lower-ranking monkeys, in a perfect linear relationship with rank. No other group member distributed social favors in such perfect accordance with rank or a tit-for-tat policy.

Wiggy was a tiny, slender adult female who had by far the sweetest temperament of anyone in the group. She would diligently groom anyone who requested grooming, whether it was the alpha male or one of the group's most despised pariahs, and she was remarkably forgiving. Tattle, the lowest-ranking female, was apparently just as dim-witted as Squint was Machiavellian. Tattle spent the vast majority of her time on the edge of the group, generally feeding in her own tree rather than squeezing into the main feeding tree with the rest of the group. Food seemed to be of far greater importance to Tattle than social contacts. However, she shamelessly pandered to the adult males whenever they came near her. Whenever they approached, she would begin threatening me and run to them for help, apparently using me as a social tool to improve her relationship with the males (hence the name Tattle).

Toward the end of our two pilot studies, the tables had turned. Though my monkeys were still a challenge to find and follow, I was too caught up in the capuchin social drama to have even a trace of the "howler envy" that I had experienced in our first weeks at Lomas Barbudal. And Joe was bored to a frenzy by the howlers, who lay around doing nothing all day while he sat below, serving as a pin cushion for the immense clouds of mosquitoes that gathered to feed on this immobile cluster of mammals. I came home day after day with tantalizing new stories of the capuchins' antics, whereas the highlight of Joe's week would be "Flora groomed Alberta for three seconds."

After we returned to Ann Arbor in August, I spent the next eight months recruiting field assistants and raising funds to support a lengthy study of Abby's group. Julie Gros-Louis, who had been Joe's field assistant (along with me) during Joe's work at Cayo Santiago and was now in Australia studying dolphins, agreed to spend the summer of 1991 helping me. I obtained enough grant money from the National Science

Foundation, National Geographic, the Leakey Foundation, Sigma Xi, and various branches of the University of Michigan to conduct a two-year study. I also struck a deal with Gordon and Jutta Frankie, the conservationists who had helped establish the Lomas Barbudal Biological Reserve. We would do volunteer work for their organization, Friends of Lomas Barbudal, in exchange for the use of a vehicle. This would enable us to live in town, where we would have electricity to charge batteries for our equipment, instead of camping in the forest as we had done for so much of the four-month pilot study. Meanwhile Joe applied for funding to do his postdoctoral project on howlers, though with a noticeable lack of enthusiasm. However, he came around gradually to the idea of serving as my field assistant, as I had done for him during his doctoral research, and he was actually relieved when his grant proposals were turned down.

The summer of 1991, during which my capuchin research began in earnest, was perhaps the happiest time of my life. Joe, Julie, and I were joined by Laura Sirot, another Cayo Santiago veteran who volunteered her time on my project. We were highly compatible as colleagues and as housemates, and the project ran virtually seamlessly—or at least it seemed to, since Julie and Laura each had such a good sense of humor that they made all of our blunders seem like part of a fun game rather than an embarrassment. We all enjoyed the high spirit of adventure, possessed a strong work ethic, and had an insatiable curiosity about the monkeys. Far from complaining about the many disgusting things that happened to us—monkey poop in the mouth, ant colonies in our shoes or pockets, accidental ingestion of insects, fungus-infested feet, festering sores, close encounters with rotting carcasses, and so on—Julie and Laura established the annual Festerfest awards, in which researchers win prizes for enduring the most disgusting or embarrassing traumas (such as the time one of us fell face first on top of a rotting, maggot-filled anteater carcass, or the time a local hairdresser, while administering a haircut, uncovered a one-centimeter-wide engorged tick firmly attached to a field assistant).

Julie and Laura were particularly fearless about plunging into the local culture, and we are grateful that they dragged us along with them. During that summer our Spanish improved immensely, and we established many lifelong friendships with local people. By the third month of the 1991 field season, we had the data collection running smoothly, and the monkeys were perfectly habituated to our presence, ignoring us as they went about their daily routines. The monkeys moved so quickly, and their behavior was so variable, that we could not collect data on coded check sheets, the way most primatologists do. We collected data as a team, with one person narrating the focal animal's behavior into a microcassette recorder and another calling out the ID's of the monkeys coming up to interact with the focal animal. Since there were four of us, the other two observers were free to roam through the capuchin group, collecting ad lib data on interesting events such as sex, predation on vertebrates, and strange play behaviors. Unfortunately Julie and Laura had to go back to the United States in August to continue their undergraduate studies.

Joe and I carried on alone from September 1991 through May 1992. It was exhausting work with just two of us, and looking back on it I can hardly believe we managed. For five days in a row we would drive out to the forest well before dawn, and we would stay until the monkeys were in their sleeping trees shortly after dusk. We stayed up far into the night doing laundry (by hand, of course), cooking, and dishwashing, and then, after a little sleep, began the grueling process again. We spent two days a week in Bagaces, transcribing data tapes and doing volunteer work for Friends of Lomas Barbudal. Much time was wasted at the beginning of each work week, trying to find the monkeys again. We decided that we needed another field assistant so that we could take turns staying home, with one person doing the housework while the other two were in the forest. After much pleading, we managed to talk Julie into joining us for another year. Once she arrived, we were able to monitor the monkeys for twenty-five consecutive days out of each month.

Although individual adult monkeys were easily recognizable early on, the juveniles and infants were more difficult to identify. A few months into the project, when the monkeys trusted us enough to come within a meter or two of our feet, we discovered that it was possible to mark the infants and younger juveniles with Clairol Born Blonde hair dye, by surreptitiously squirting their tails and rumps with a syringe as they walked past us. Once we could identify the youngsters as well as the adults, the social dynamics became even more fascinating. Once the infants were marked, we realized that this species engaged in an unusual degree of alloparenting, or "babysitting." The younger infants traveled in a pack,

Susan watches Abby's group during her early fieldwork, as Nanny grooms Curmudgeon. Photo: Joseph H. Manson.

and if their mother was not available, they demanded milk from any adult female who was nearby. They also accepted frequent rides from various group members, including adult males. The subordinate adult males, such as Guapo, formed quite close attachments to particular infant males.

In May 1993 we returned to Ann Arbor, our carry-on luggage filled almost entirely with hundreds of microcassettes containing our irreplaceable data. As I undertook the onerous job of transcribing the tapes—a task that took me eighteen months to complete, despite the fact that I worked at it all day long, seven days a week—I often found myself wondering what the characters in Abby's group were up to back in Lomas Barbudal.

Even before I began to analyze the data, it was clear to me that capuchins exhibited a level of social complexity comparable to that of Old World monkeys, and therefore that the answer to my original research question—Do capuchins' large brains falsify the social intelligence hypothesis?—was no. By the time I finished my initial project in 1995, I had validated that intuition with a large set of data analyses. There was no doubt that these monkeys had strongly differentiated social relationships. The females had a clearly defined, linear dominance hierarchy, and the quality of their social relationships remained constant over long periods of time.[13] From 1990 through 1993, the only substantive change observed in the quality of female relationships concerned the ascent of Diablita, an adolescent female, up the dominance hierarchy. Male relationships were more volatile, and females' preferences for male social partners seemed to depend more on dominance rank than on other personal characteristics of the males.[14] Dominance rank also affected grooming patterns, with higher-ranking females receiving more grooming than lower-ranking monkeys. The smaller the difference between two females with regard to dominance rank, the more they groomed each other.

The monkeys also seemed to exchange social favors, indicating that they were keeping track of who did what for whom. Females formed coalitions preferentially with the females who groomed them most, although this finding could also be explained in part by the fact that higher-ranking females both received more grooming and were more willing to participate in coalitions in general. Males provided coalitionary aid most frequently to those females who groomed them the most, at least during Paul Bunyan's term as alpha male following a rank reversal in 1992 (see Chapter 7). Also, males and females seemed to exchange coalitionary aid during this same time period.

The prevalence of coalitionary aggression in capuchins made their political landscape much more complex than that of most neotropical primates, and it seemed logical that the frequency of such triadic interactions would select for a mental capacity for triadic awareness—in other words, an ability to monitor and remember the quality of interactions observed between other group members. I paid careful attention in my initial months of research to the circumstances surrounding coalition formation in the capuchins, noting who was standing around when fights broke out, who took whose side, and who solicited aid from whom. Requests for coalitionary aid were easy to observe because they usually consisted of a species-typical gesture, the "headflag," in which the initiator jerkily alternates his or her gaze between an opponent and the hoped-for supporter.

In 2001 Joe and I used these data to answer another research question: What decision rules do capuchins use when they ask each other for coalitional support? First, for each pair of monkeys in the group, we calculated the proportion of their interactions that were friendly or cooperative (e.g., grooming, cuddling, playing, or coalition forming) as opposed to agonistic (e.g., fighting, food stealing, or fearful behaviors). We chose this more complex index of affiliation rather than simply looking at grooming rates or frequency of proximity, the most common measures in primatology, because members of different capuchin age-sex classes have strikingly different ways of expressing their affiliative tendencies. Males, for example, hardly ever groom each other the way females do, but they often play or rest in contact with one another.

Knowing all the monkeys' dominance ranks, and having ascertained the quality of their social relationships with one another, we persuaded our colleague Clark Barrett at UCLA to write a computer program that would select a monkey at random from the audience for each fight, and then check to see whether that decision/selection conformed to the decision rule we were testing. By running this program 20,000 times, we could figure out the distribution of proportions of times the monkeys would be expected to follow any particular decision rule if they were soliciting audience members at random, and see whether the actual decisions made were statistically more likely to be in line with the decision rule.[15] We found that the monkeys were significantly more likely to solicit aid from a monkey who outranked their opponent rather than one that was subordinate to their opponent. However, they also chose the highest-ranking audience member more often than would be expected by chance, so it was difficult to know for certain whether the monkeys were exhibiting a triadic understanding of rank or just choosing the toughest-looking capuchin in the crowd. The results regarding relationship quality were clearer, however. Monkeys solicited aid significantly more often from a monkey with whom they were better friends than their opponent was. This indicated an understanding of other group members' relationships that couldn't be explained by simpler rules. For example, we checked to see whether they could coincidentally produce this result simply by picking their "best friend" from the audience (since on average their best friend would probably be a better friend of theirs than of their opponent), without having any good understanding of the relationship between the would-be supporter and their opponent. However, they did not tend to pick their best friend.

Clearly the monkeys had a sophisticated understanding of group dynamics that they used in a strategic manner. But did they have a theory of mind, or were they merely good at detecting regularities in the behavioral patterns of their groupmates? The best work done on these issues in Old World monkeys had been carried out by Robert Seyfarth and Dorothy Cheney, in a study of vervet monkeys.[8] They concluded that monkeys such as vervets and macaques were superb natural historians of social dynamics but poor mind readers, and it seemed to us likely that capuchins would follow that pattern.

Theory of mind is an elusive concept to define and measure, particularly for someone working in the field, where conditions cannot be controlled. The best I could do was to be on the lookout for possible instances of tactical deception of the sort included in Whiten and Byrne's cross-species survey.[10] I did on very rare occasions see what seemed to be false alarm calls—instances in which a monkey who was in big trouble with the rest of the group would suddenly begin producing snake alarm calls while staring urgently at the ground. I always made a point of looking for the predator, so that I could match my recordings with the proper stimulus that elicited the call. Plus, as a practical matter, I preferred not to step on poisonous snakes. Sometimes there was no visible predator where the monkey was calling. In some cases, of course, there were holes in the ground that a predator could have slipped into before I arrived on the scene. But in other cases, I removed all leaf litter and found neither holes nor predators.

The most memorable of these instances took place on August 26, 1992, when the subordinate male Guapo was being pursued by a coalition consisting of the scariest members of the group: Curmudgeon, Abby, and Paul (that is, the three highest-ranking members aside from himself). Guapo suddenly stopped in his tracks and began to produce frantic snake alarm calls while looking at the ground. I was standing by him and could plainly see that there was nothing there but bare ground. He headflagged to Curmudgeon for support against the imaginary snake.

Guapo's pursuers stopped short and stood up on their hind legs to see if there was a snake. After a cautious inspection, they once again began threatening Guapo. Switching tactics, he glanced up at a passing magpie jay (a nonmenacing bird) and did three bird alarm calls in rapid succession—calls that are usually reserved for large raptors and owls. Guapo's opponents looked up, saw that it was not a dangerous bird, and again resumed threatening Guapo. He reverted to the snake alarm call tactic, once again vehemently bouncing at the bare patch of ground, threatening the "snake" vocally, and insistently headflagging to Curmudgeon for aid eleven more times. Although Curmudgeon continued to glare at Guapo for a bit longer, the rest of the group stopped threatening him, and he was able to resume foraging for insects, moving slowly and nonchalantly toward Curmudgeon while occasionally casting a furtive glance in his direction.

Observations such as this one, and the incident of the juvenile baboon tricking his mother into chasing an adult female away from a valued food item, suggest a capacity for reasoning about others' beliefs, but they can also be explained by cognitively simpler processes. Did Guapo know that he was tricking his antagonists into thinking there was a snake there when really there wasn't? Or had he merely learned from past incidents that when he produced snake alarms, his antagonists stopped bothering him? Perhaps an even simpler explanation could suffice: maybe Guapo was so upset at being chased by this formidable coalition that he was imagining danger everywhere. In the absence of experiments, it is difficult to distinguish between alternative interpretations.

In recent years, psychologists working with captive animals have performed cognitive experiments on brown capuchins to investigate what these monkeys know about others' knowledge or intentions, and whether they can use this information to deceive others. The results have been mixed. Brian Hare and his colleagues found that when two monkeys were simultaneously released into a cage containing two food items, one of which had been visible to both animals and the other visible only to the subordinate, the subordinate preferred to approach the latter.[16] It was not clear whether this was because the subordinate knew that the dominant monkey could not see that food, or because the subordinate was using subtle cues from the dominant about its intention to approach the more visible food, and wished to avoid competition.

Knowing what another animal can and cannot see is interpreted by some researchers as being a form of knowledge about the contents of its mind or, at least, as a cue useful for predicting its subsequent behavior. However, the subordinates in the experiment may have been using subtle "intention movements" by dominants, rather than eye gaze, as their source of information, because when subordinates were released into the cage slightly before the dominant monkeys, their preference for the hidden food vanished.

In another experiment, a team of Japanese researchers created an apparatus in which a subordinate brown capuchin chose to open one of two boxes, one containing food and the other empty, that lay between himself and a dominant monkey.[17] Only the subordinate could see which of the two had food in it, and only the subordinate could open the boxes, yet the dominant could reach into the boxes as easily as the subordinate. One of the four monkeys tested showed some tendency to open the empty box first, suggesting that it was tricking the dominant into groping about in the empty box while it grabbed the food from the second box. However, this subordinate did not in fact move quickly toward the box containing the food. Moreover, opening the empty box first did not increase the amount of food acquired by subordinates.

Finally, Robert Mitchell and James Anderson investigated whether brown capuchins could deceive a human trainer as a means to obtain a reward.[18] Each monkey was shown the location of a food item in one of two containers, after which a trainer who had not seen the baiting approached. If the monkey indicated which container held the food, the trainer playing the "cooperative" role would give the food to the monkey, whereas the trainer playing the "uncooperative" role would take the food and eat it. If the monkey directed the uncooperative trainer to the empty container, then the cooperative trainer would approach and give the monkey the food. One of the study's two capuchin subjects systematically directed the uncooperative trainer to the wrong container. However, as in the reports of deceptive incidents from the wild, this behavior could have been the result of a simple learned contingency ("If the uncooperative trainer arrives, indicate the wrong container to get the food reward") rather than evidence for a theory of mind and tactical deception.

Thus, even so many years after the start of my project, the extent to which capuchins exhibit a theory of mind and a proficiency at tactical deception is still unclear to us; this remains true for most Old World primates as well. But it is obvious that white-faced capuchins exhibit a social complexity that rivals that of the Old World primates, as well as a sophisticated understanding of groupmates' relationships with one another. Clearly capuchins are no social morons. Anyone seeking to knock down the social-intelligence hypothesis would have to look elsewhere for evidence. However, just because capuchins are socially astute does not mean that the alternative hypotheses for the evolution of intelligence need be discounted. Most likely, multiple selective pressures have contributed to the evolution of intelligence and large brains. Indeed, both the spatial-memory and extractive-foraging hypotheses have recently received additional support from studies of free-living brown capuchins. Charles Janson and colleagues, using artificial feeding platforms during periods of fruit scarcity, found that these monkeys follow the most efficient travel routes between food sources.[19] Eduardo Ottoni and his colleagues have observed brown capuchins cracking nuts with stone hammers, exactly as chimpanzees in Ivory Coast are famous for doing.[20]

The more I learned about capuchins, the more fascinated I was by the parallels between them and certain Old World primate species. Like the more commonly studied Old World monkeys, they have a "female-bonded" society, with females remaining in their natal groups and forming stable, affiliative relationships with other females. Although this arrangement is common in Old World monkeys, it has apparently evolved independently in capuchins and common squirrel monkeys, since the opposite pattern, female dispersal, is the norm among New World monkeys. But were the "nuts and bolts" of female bonding the same in capuchins as in Old World monkeys, with females nepotistically aiding their own offspring and siblings to attain a high dominance rank? This was a question that merited future research. Perhaps capuchins could give us insight into the evolutionary reasons for female bonding.

I also saw many intriguing parallels between capuchins, on the one hand, and chimpanzees and humans, on the other. Coalitions are integral to all three societies, in that individuals must exhibit a detailed understanding of others' relationships in order to be successful politicians. While coalitions are found in many Old World primate species, capuchins are virtually unique among New World primates in their capacity to form complex alliances, so in this respect they represent an evolutionary convergence with humans and many other Old World primates. Cooperation is important at many levels: for defending or gaining access to mates, for defending offspring, for cooperatively raising infants, for gaining access to food, and for warding off predators. Because cooperation is so crucial to the monkeys, they have developed a complex repertoire of signals for communicating about their relationships that includes many sexual elements. I became fascinated by these signals and wanted to work out the specific meanings of each one. There was very little in the theoretical literature on how animals communicate about their social relationships. I wanted to better understand what "design features" of signals were appropriate for conveying particular meanings and intents. Capuchins, like chimps, can be extremely emotional and hot-tempered animals, and their conflicts can be deadly when unchecked. As I was to discover later in the study, the main cause of mortality in capuchins is conflicts with other capuchins. So there is excellent motivation for the animals to develop a means of effectively negotiating aspects of their relationships—before the stakes get too high.

The parallels with chimpanzees and humans are not confined to social relationships, however. Capuchins, like chimps, are omnivorous and include vertebrate prey in their diet. They are remarkably pugnacious and stubborn animals, as is evidenced in their strategies both for obtaining food and for deterring predators. It seemed to me that capuchins' interactions with the physical world were likely to be as interesting as their social lives. They showed parallels with humans, again, regarding their capacity for tool use and other foraging-related problem solving, their tendency to share food, and their enthusiasm for hunting. More and more questions arose that could not be answered with my first data set. Do capuchins hunt cooperatively? How do they acquire their food-processing skills: do they learn by trial and error or by observing fellow group members? Do all capuchin populations solve their subsistence problems in the same way, or are there cultural traditions unique to particular cliques or social groups, as there are in humans? Obviously I was going to have to spend many more years in the field to answer all of these questions.

In thinking about the monkeys' many parallels with our hominid ancestors, I began to realize the value of using capuchins as a model for investigating the evolution of these traits that are considered to be so quintessentially human. Chimpanzees are currently the "model system" most typically used to explain what our hominid ancestors were like, because they share many of the traits anthropologists would like to explain in humans. Chimpanzees are so closely related to humans that we no doubt share many traits with them owing simply to "phylogenetic inertia"—that is, because we both inherited various traits from a common ancestor, who evolved those traits in response to a particular adaptive challenge in its environment. But traits may persist longer than they are useful, so long as they do not impose a big cost on the organism. So whereas comparing the similarities between chimps and humans can tell us a lot about what our common ancestors were like, studies of chimps cannot tell us much about *why* our ancestors had big brains, shared their food, and formed complex alliances. To know that, it is essential to find instances in which other species, more distantly related to humans, have independently evolved the traits of interest. We can then test hypotheses about the functions of these traits by looking for the adaptive challenges that distinguish the species that have evolved them from related species that have not evolved them. These challenges could include features of the environment, such as food sources and predators, or perhaps preexisting traits of the species itself, or most likely some distinct combination of factors. We have a long way to go before reaching a convincing answer to the question of why we became the creatures we are, but data about capuchin monkeys will undoubtedly play a major role in getting us there.

# CHAPTER 3

# The Challenges of Foraging and Self-Medication

Although I had come to Lomas to study the social interactions of the capuchins, it was difficult not to become absorbed in the details of their foraging and other interactions with their physical environment. In contrast to the howler monkey diet, which consists of leaves and fruits that can be popped into the mouth without processing, or even without using the hands in most cases, the capuchin diet is largely made up of items that have to be bludgeoned, torn apart, disarmed, or killed. Capuchins love a good challenge, and even if they have just gorged themselves on fruit they will rarely pass up an opportunity to rip apart a dead branch in search of insects, investigate a tree hole, or harass an animal that might be good to eat, no matter how dangerous it may be.

As I watched the monkeys forage, I became increasingly aware of the cognitive problems that their diet poses for them, and the selective pressure that these challenges exert on their cognition. They need to solve spatial problems to find their food, defeat the mechanical defenses of plants, and outwit animal prey. For a capuchin, mealtime is as full of high drama as social time. And as I was to find out much later, the monkeys' social relationships play a role in the way food is acquired and processed, owing to the influence of social learning. Extractive foraging tasks that require multiple steps pose special challenges, and in these instances, it is useful to pick up helpful tips from more experienced group members. Meat is a rare and highly valued item in the capuchin diet, and it can be dangerous to obtain it. Hence, hunting by adults attracts a great deal of interest from other group members who want to share in the spoils as well as to learn more about how to obtain meat themselves.

It is Easter Sunday, 1993, and Joe and I are with Abby's group right by the *poza*, the swimming hole that is the focal point for the ecotourism business at the northern edge of the reserve. It is still fairly early in the morning, and Joe and I are the only humans in the forest, enjoying the peace and quiet. Suddenly we hear the shrill undulating squeal of a coati. By now this had become an all too familiar cry—it was the sound made by a coati pup as it is being eaten alive by a capuchin. We dash toward the sound, and as we run we see an adult female coati rushing to her pup's rescue. Several monkeys are charging toward the sound as well. When we arrive, we see Wiggy, a sweet subordinate adult female, holding the captured pup. She is being shadowed by her yearling son Yoyo and by the juvenile male Quizzler. Wiggy pauses to rip open the pup's throat, producing the tonal peeping vocalizations capuchins generally make while foraging. She is obviously nervous about the keen interest of Quizzler and keeps moving away from the growing crowd of beggars, which now numbers four monkeys. After eating only a third of the pup, Wiggy drops it, and three of the beggars scramble to the ground to retrieve the remains. Meanwhile the adult female Nanny has resumed the raid on the coatis and is lunging at the mother coati, who is now sitting on her nest to protect her remaining pups.

Across the river we hear the sound of a second nest raid. Often members of the same coati group will build their nests near one another. We can hear the screams of two pups—Tattle has one, and the second pup is dropped to the ground in a skirmish between two monkeys. Hobbes (the nastiest, shortest, and most brutish of the juvenile males) grabs the dropped pup and futilely attempts to process the flailing coati by scrubbing its hide against a branch to remove the hairs, the way he would process a hairy *Sloanea* fruit. Tattle bites into her pup with relish, and soon her chubby white cheeks are smeared with rougelike circles of blood, giving her the appearance of a macabre clown. Squint emerges from the vine tangle containing the second nest, holding another screaming pup.

All the carcass owners are being pursued by other monkeys, who approach and sniff the pup and the owner's mouth, and tentatively touch the carcass while looking at the owner's face. Squint's yearling daughter Broma reaches into the abdomen of Squint's carcass and grasps a piece of intestine. Then she steps back out of mom's reach and starts to reel in the guts, pausing to nibble them every now and then. It is rare for a monkey to steal food outright, but it is very common, especially with highly prized items such as meat, for them to gently lick or bite off a piece of another's food item. This is generally tolerated, but if the beggar becomes too demanding, the food owner may gently nip or push the beggar. Not surprisingly, females are more tolerant of begging from their own offspring than from less closely related individuals. Some beggars use the amusing tactic of hopping on the food owner's back to form a coalitionary posture against the carcass. After bouncing and performing some squeaky threats against the helpless victim (who is often already dead), they ask for a portion of the meat in exchange for their "help," and sometimes this odd strategy proves successful.

Joe and I are kept very busy during coati nest raids, because we are interested in documenting both the hunting techniques and the patterns of food sharing and theft. Our strategy is to do "carcass follows"—to determine the fate of each pup and see which portions of each carcass are devoured by each monkey, and also to record the social interactions between food owners and beggars. Our job is made easier (as well as more traumatic and disgusting) by the fact that the pups typically scream for the first twenty to thirty minutes of the process, which makes it easy for us to keep track of their locations. Coati pups are so undeveloped at birth that the monkeys feel no need to kill them before eating them. The pups scream but do not actively resist. Most nest raids yield only four pups, so we each generally follow the fates of two. But on this day, with two simultaneous nest raids to document, we are having a tough time keeping track of all the monkeys' victims.

In the midst of this bloody mayhem, a busload of tourists arrives at the *poza*, decked out with cameras and video recorders. Fortunately the pups had all died in the preceding few minutes, so there is no more screaming. I silently curse my luck, though. I am all in favor of ecotourism, and environmental education in general, but it is always frustrating to be interrupted in the midst of documenting an interesting event in order to answer people's questions. And I am certain that the tourists will have a *lot* of questions about this particular event. Coati infants, with their soulful brown eyes, long curvy noses, and graceful ringed tails, could well be the poster children for the tropical forest. In fact, Toño Pizote (Tony Coati) is the Smoky the Bear of the Costa Rican Park Service, and many vehicles proudly display a picture of him admonishing farmers to protect dry forests and follow government regulations regarding fire control. When you couple the symbolic importance of the coati with the cuteness and helplessness of the pups, you can imagine how unpopular capuchin monkeys would become once it became generally known that they ruthlessly slaughter the pups right in front of their traumatized mothers.

As the tourists close in on me, I hear them uttering the typical tourist lines:

"Oh, Fred, just *look* at that, will you?!! Are those monkeys? My, they are cute. Get the camera ready. Charlene will be so jealous when she sees these shots."

"Look, there is another one! I see six! No, seven!"

"Where? Oh there! And here is another one, did you see that one?"

"Here, kitty kitty kitty! Look this way! What are they?"

"I think they are some of those howler monkeys we heard yesterday."

(Tarzan yell from an adolescent boy.)

"What are those red fruits they are eating?"

"Oh, look, Herb, they are sharing the fruit, how sweet. Get the binoculars out, will you?"

I cringe, wondering how long it will be before they realize the red fruit is a bloody infant.

"Oh my!! Oh dear. I'm not sure that *is* a fruit, Herb. It looks like . . . like an animal or something. But monkeys don't eat meat do they? I thought they were peaceful vegetarians. Oh, this is dreadful. I can't look. Here, you take the binoculars."

The shocking news passes from mouth to mouth, and soon people are staring at me suspiciously. As I struggle to finish collecting the data, talking into my tape recorder, I can hear the tourists forming various hypotheses about what is happening. Eventually they reach the consensus that I am an evil scientist who has given the monkeys meat as part of an awful experiment. Obviously it is time to stop collecting data and start working to clear my name. So I radio Joe to tell him that I must abandon my carcass follow and ask him to carry on as best he can, picking up the action on my side of the river as soon as he has finished documenting the fates of the victims on his side.

Given our own species' carnivorous history, it is amazing how reluctant many Westerners are to accept the possibility that many of our closest relatives eat meat as well. Most people are willing to concede that felines' and canines' carnivorous diets are natural and therefore acceptable, although even some dog owners force their pets into vegetarianism. But people are much more squeamish about predation by primates. Forty years ago, Jane Goodall's reports of hunting by chimpanzees shocked people's sensibilities as much as they surprised scientists. Perhaps omnivores, unlike true carnivores, are seen to have the option of eliminating meat from their diet. When monkeys eat meat, it is often assumed that this is because humans have degraded their environment or somehow forced them to do it. The former hypothesis was, in fact, the not-too-subtle subtext of an otherwise excellent TV documentary produced in 1992 about the white-faced capuchins at a site not far from Lomas Barbudal. Mother Nature could not possibly have intended cute animals like monkeys to devour other cute animals.

Apprehensively, I turn to introduce myself to the throng of tourists and begin to explain that coati pups are a natural part of the monkeys' diet. I know from my colleague Lisa Rose at the Santa Rosa research site that monkeys regularly consume coatis there as well, and there are reports of coati predation from a third site.[1] At Santa Rosa, which is about fifty kilometers northwest of Lomas, there are researchers who monitor the coati population as well as the monkey population. Even though the monkeys seem to kill an astounding portion of the coati-pup population each year, there is no good evidence that the coati population is declining or that the capuchins' inclusion of coatis in the diet is a recent event in their evolutionary history, coinciding with the increase in the human population. I then appease the shocked tourists by telling them about some of the nicer aspects of capuchin social life, such as babysitting and their loyalty to their friends. After an hour or so, they leave, satisfied with their monkey encounter, and I am able to go back to work. I catch up with the monkeys, who have brought their "Easter Sunday Massacre" to a close.

Although I was as horrified as the tourists the first time I saw a coati predation episode, I eventually became accustomed to seeing matters from the monkey perspective. When you spend all of your waking hours following capuchins and trying to imagine what is going on inside their heads, you soon start viewing even the cutest critters as just another form of monkey chow, and steeling yourself against feeling much empathy for the prey. Even after years of seeing these incidents, however, I still feel a bit sick when I see the mother coati returning to her nest and desperately doing battle against the monkeys in a futile attempt to save her pups. Although coatis are about twice the size of capuchins and have formidable jaws, they are handicapped by their lesser agility and exceedingly poor eyesight. Following her nose and the sound of the screaming pup, a coati will clumsily climb a vine to try to reach the pup and the attacker. The capuchin monkey will keep eating, unconcerned, and blithely hop to another vine or branch when the mother coati is just about a foot away, leaving her to sniff the branch where it had been sitting, in a state of extreme confusion and agitation.

Among my research teams, one of the most spectacular Festerfest awards of all time was presented to a vegetarian field assistant, Kathryn Atkins, for her experience during a coati nest raid in 1997. A highly empathetic person, Kathryn had been dreading coati season all year, but she was doing quite well at keeping her cool and maintaining her usual high standard of excellence in data collection during these predation incidents, despite the fact that they disgusted her intensely. We had been watching Feo, a decrepit old male in Rambo's group, eat the remains of the last pup from a litter, and the rest of the capuchins had completely lost interest and moved on to other activities. It had started to rain, so we put on our rain ponchos and terminated the follow of Feo, unable to see the necessary details in the downpour. After we had been standing there for a while waiting for the rain to pass, I felt a hard object hit my back. Disconcertingly, I did not hear a second thump as it hit the ground. I wondered what it was and why it was stuck to my back. I asked Kathryn to take a look. After one glance, she screamed, and squealed, "I *hate* this job!!!" I was rather taken aback, since she was normally cheerful, unflappable, and totally dedicated to the work; this response seemed totally out of character for her. "What's wrong?" I asked nervously. Kathryn gritted her teeth and said, "Believe me, you don't want to know. Just hold still, will you?" She grabbed a stick from the ground and started prodding my back with it. Finally I was able to wheedle the information

The five-year-old male Cassie devours a squirrel pup. Photo: Hannah Gilkenson.

out of her. Feo had dropped the slimy, bloody, saliva-sodden head of his coati pup into the hood of my rain poncho. Despite her extreme aversion to the situation, Kathryn had valiantly removed this nasty object from my hood before I made the mistake of pulling it up over my head. After laughing helplessly for several minutes, I thanked her profusely and gave her the rest of the day off to recuperate from her ordeal.

Although the vertebrate predation events make a big impression on observers, in part because the capuchins get so excited about meat, vertebrates form only a small portion of the capuchin diet during most months (representing about 1 percent of their foraging time). Roughly 45 percent of the monkeys' foraging time is spent eating or processing fruit, while 47 percent is spent searching for or consuming insects and other invertebrates; the remainder of the diet is pith, flowers, and verte-brates.[2] Watching capuchins, you get the impression that they are constantly vigilant for a foraging opportunity. Even when they are absorbed in grooming or in coalitionary aggression, their eyes will sometimes dart to the side and a hand will flash into the air to grasp a flying insect that will then be popped into the mouth and chewed as the social activity is resumed. As the monkeys walk through the trees, they constantly pause to tap sticks, stick their hands into tree holes, pry up loose bark, and investigate clusters of dead leaves. Many species of ants occupy the hollow insides of twigs, and the monkeys presumably tap the sticks to coax out the ants. Not all capuchins tap sticks—this is a trait practiced by only 80 percent of the monkeys in the population. If a stick seems like a likely foraging prospect, the monkey will break it off and use his incisor teeth to split it down the middle. Then, with his tongue darting in and out of the center of the stick, the monkey picks up the ants and spits out "sawdust."

The capuchins also forage on the ants that protect acacia trees. Acacias and their resident ant colonies are famous examples of symbiosis, the mutual dependence of two species. The trees provide the ants with food and with thorns in which to house themselves, and the ants use their painful bite to protect the tree from leaf-eating animals. Hanging by his tail from a tree above an acacia, a capuchin will reach down to pluck off the thorns housing the ant larvae. When the adult ants swarm out and begin biting the monkey, he deftly wipes them off with his other hand or pops them into his mouth, chewing quickly to kill them and then wiping his hands on a branch, presumably because of the itching bites.

One species of tree-hole-dwelling ant is large and unusually fast, with a particularly nasty bite. When the monkeys discover these ants, they move into high gear and start frantically popping them into their mouths and crunching at top speed, in a race to devour the angrily swarming ants before they get bitten in return. And of course the monkeys also raid the bird nests that are built in protective acacia trees.

Stinkbugs are another staple of the capuchin diet. These insects produce an acidic substance that burns the skin and fills the air with a foul aroma. Although my intrepid field assistants have sampled most of the capuchin foods, this is one that no one has been tempted to taste; the acid burns the skin so badly that I shudder to think how it would feel on the tongue. Nonetheless, the monkeys consume them with great gusto from an early age. A few monkeys in the population even crush stinkbugs and rub them all over their bodies before consuming them.

From the middle to the end of the rainy season, the monkeys' favorite insect prey is paper wasp (*Polistes instabilis*) larvae. Capuchins are major predators on this species. One rainy season I calculated that each monkey consumed, on average, 1.6 wasp nests per hour, such that Abby's group ate about 400 nests per day. Paper wasps build nests in the lower canopy, inconveniently located somewhere between eye level and knee level for a primatologist, and they tend to conceal them in dense foliage. Paper wasps are not particularly aggressive, fortunately. Typically four to ten wasps will be standing on their open-celled nest tending the larvae, and they stare balefully at any approaching predator. They do not react until the invader gets within a few inches of them, whereupon they start quivering and fidgeting and finally leave the nest, swooping out to deliver their silent stings. They will chase the predator for only a few meters before returning to the nest.

Although the stings are extremely painful, the worst aspect of paper wasp season is not the stings themselves but the anticipation of being stung. As we throw our bodies into the dense undergrowth, forging a path in order to follow the foraging monkeys, we know with certainty that at some point very soon we're going to disturb a concealed wasp nest and receive multiple stings. And for a few seconds after being stung, we'll have that moment of terror when we think, "Killer bees!! And here I am, trapped in a thorny vine tangle!" before we realize with great relief that they are just paper wasps. Many a field assistant has resigned because of the psychological exhaustion from "wasp anticipation." No one on my research team has ever tried eating paper wasp larvae, but we have had enough experience with the stings to be certain that, from a human perspective, they would not be worth the pain and suffering.

The monkeys are not as easily deterred as humans, of course. Most do try to avoid the stings, but a few adult individuals are so pain-resistant and greedy that they will consume a wasp nest on the spot, without running away, even though the wasps sting them again and again. Infants who are not yet skilled enough to plan a successful hit-and-run attack will tough it out through many stings in order to feast on the wasp larvae. It is amusing to watch these first attempts: the infant reaches cautiously toward the larvae-filled nest and is stung by several wasps. Flinching, the monkey withdraws his stung hand, screws up his eyes, and glowers at the nest. This routine is repeated several times until the frustrated infant finally closes his eyes, grabs the nest, pulls with all his strength to tear it from the branch, and zooms away with it, frantically swatting wasps off his back.

The vast majority of capuchins become highly skilled at wasp-nest capture at an early age. It is fun to watch them strategize. When a monkey spots a wasp nest, you can almost see the gears turning in his head as his gaze strays up and down the branch of the plant to which it is attached, and also to other plants in the vicinity that will offer a good escape route. If possible, the monkey approaches the nest via another plant so that the plant bearing the nest will not be disturbed until the last possible moment. Then the monkey zips his hand in, quickly grabs the nest, and runs like crazy, leaving the confused wasps flying about in a cloud around the substrate that had held the nest. The monkeys typically pluck the developing paper-wasp larvae out of the cells and discard the nest itself. However, with some species of wasps they eat the nests in their entirety. For example, *Polybia occidentalis* nests contain some honey, and a monkey is capable of devouring a *Polybia* nest that is larger than its own head.

The most beautiful and, perhaps, spectacular type of wasp nest in the forest is the warrior wasp nest. These nests, which can be two feet long, are typically built on the smooth underside of *Bursera* tree trunks. The cells are covered in a tan paperlike covering that is composed of artistically patterned ridges. At the top of the nest is a round "spout" that serves as a doorway for the enormous violet-black wasps that dwell inside. Warrior wasps are peaceful when foraging, but when disturbed, the hundreds of wasps dwelling inside the colony will first drum their feet on the nest walls and then swarm out in droves. They are famous for having an extraordinarily painful sting.

Monkeys normally do not mess with warrior wasp nests, but they seem well aware of the potential danger, and occasionally I have seen boister-ous, hormone-crazed young adult males launch kamikaze attacks on these nests, perhaps in an attempt to show off for the rest of the group. In one memorable instance, Guapo walked along the top of a *Bursera* tree that was growing at an angle and glared down at a nest. He crept up to the nest, bounced cautiously, lunged forward a couple of times, like a child trying to decide whether he has the nerve to dive off a high div-ing board, and then gave the nest a hard swat. Without waiting to see what would happen, he ran as fast as he could, out of the tree and down to the ground. The wasps swarmed out in droves and pursued him as he ran straight toward me. I escaped to a safe distance and watched as Guapo dashed off, scratching and swatting away wasps. He did not dare to come back and approach the fallen nest, which now lay smashed on the ground, and so he did not get anything to eat as a reward for his act of bravado.

When the rains come at the end of the dry season, caterpillar season commences, and the monkeys feast on their juicy flesh for hours on end each day. Because particular caterpillar species specialize on particular tree species, caterpillars tend to be as patchily distributed as fruits, and when they find them, a group of monkeys can happily gorge themselves on the caterpillars within a single tree for long periods of time. Some of the species they eat regularly are so large that they are the size of a capu-chin's forearm. Small caterpillars are consumed whole, but the large ones are processed specially, to remove the guts. Monkeys neatly bite off the head and give the body a quick flick of the wrist, sending the guts fly-ing out in a stream toward the ground. These caterpillar guts leave per-manent stains on clothes and also badly burn the skin of primatologists standing below the feeding sessions, so it seems likely that the gut con-tents are full of unpleasant plant toxins.

My least favorite type of caterpillar is an *Automeris* species that is cov-ered with horrendous hairs that cause painful, itchy welts, sometimes lasting for days. This green caterpillar is hard to see because it blends in with the vegetation, and when you accidentally brush your hand against one, you receive a shock of unbelievable pain—it feels as if someone has just hammered a nail clear through your hand and out the other side.

Capuchin monkeys are not nearly as wimpy as humans; they actually eat these caterpillars, though they rub their hands on a branch afterward, clearly attempting to rub away the pain in their palms. A few monkeys have figured out some neat tricks to circumvent the caterpillar's defenses. When they see an *Automeris* caterpillar, they pluck a leaf off the tree and carefully wrap it around the caterpillar. Then, touching only the leaf, they roll the wrapped caterpillar against a branch to scrub off all the pain-inducing hairs. Others apply a stick to the caterpillar, using it as a sort of rolling pin to remove the hairs. These are the only cases of true tool use that I have seen in the Lomas Barbudal capuchins.

True tool use is generally defined as the manipulation of one detached object to change the position or condition of another object.[3] Given the high rate of tool use in the brown capuchin, *Cebus apella*, it is a bit surprising that there is not more tool use in *Cebus capucinus*.[4] Brown capuchins have a rich and geographically variable tool use repertoire that is roughly comparable to that of the common chimpanzee.[5] They use hammers and anvils, sponges, cups, probes, crude stone blades, and stone digging tools.

I have often wondered why tool use is so much rarer in white-faced capuchins than in brown capuchins. It may have something to do with the nature of the foods available. In general, the most avid tool-using populations among brown capuchins live in rather marginal habitats, where some of the most prized foods (nuts, tubers, and oysters, for example) cannot be accessed effectively without a tool. The monkeys of Lomas seem to cope quite well without tools, and to the extent that "necessity is the mother of invention," this may explain why tool use is so uncommon for them. Although they could no doubt mitigate the pain of processing mechanically defended foods by more often using tools such as leaf wraps, this would probably reduce their foraging efficiency slightly, since it takes time to make or find a tool.

In the capuchins' range around Lomas Barbudal, I can think of only one fruit that is edible and also inaccessible without a tool. That is the jicaro fruit, a large gourdlike fruit a bit bigger than a monkey's head that grows in the middle of pastures and is eaten primarily by cattle. The only monkey in our population who regularly frequented pastures was Abby, when she was so old and arthritic that she could not climb trees very readily. She would venture into the shade of the clusters of mango and oak trees in the middle of a pasture and eat the fallen mangoes and acorns, free from the competition of her groupmates. Since monkeys rarely venture into the middle of pastures, it is difficult to say whether jicaro is ignored as a food source because of the processing challenge it presents, or because getting access to it would require a large detour from the monkeys' typical travel routes. Although monkeys do on occasion cross pastures, they always seem nervous doing so, perhaps because of the high density of rattlesnakes hidden in the tall grass, and also because of the dearth of trees that would offer an escape from terrestrial predators such as hunting dogs.

Although the monkeys of Lomas do not need tools to eat any item in their diet, there are some items they eat that we human observers cannot open without the aid of a machete. The panama fruit, *Sterculia apetala*, is one of these. This fruit consists of a row of very large seeds coated with a thin pulp, reminiscent of lychee fruit. The fruits are encased in a thick, *empanada*-shaped woody shell that will eventually pop open to expose them when they are fully developed and ready to be dispersed. The fruits themselves, as well as the lining of the case, are coated in tiny stinging hairs that are extraordinarily irritating. As is the case with many of the fruits in the capuchin diet, the monkeys exploit the fruit just before it is ready to open, thereby beating the competition with the regular dispersers. Capuchins are the only animal in the forest capable of prying open the hard shells, and it requires years of practice for a capuchin to perform this task efficiently. Jackson, an adult male who migrated into Abby's group in 2002, is a champion at panama processing and can open one of these fruits in approximately fifteen seconds. In contrast, some group members will manipulate a panama fruit for nearly an hour and still not be entirely successful.

Many intriguing experiments have been done by Elisabetta Visalberghi and her colleagues on the closely related *Cebus apella*, relating to their understanding of the physical world.[5,6] Although capuchins eventually do manage to solve complex foraging problems, the manner in which they accomplish these tasks does not exactly inspire confidence in their intellect. Visalberghi gave her monkeys complicated tasks, such as removing a peanut from a glass tube with a trap in the bottom of it. To obtain the peanut, the capuchins had to poke a stick of the appropriate length through the correct side, so as to push the peanut out the other end of the tube rather than into the trap. It took the monkeys quite a bit of time to solve this novel problem, and even the smartest of the monkeys seemed to employ illogical "superstitions" and rules of thumb to solve the task (based on what had worked in past trials), rather than basing their efforts on a sophisticated understanding of the properties of the objects they were dealing with.

The secret to capuchins' success as extractive foragers seems to have as much to do with their personality as with their cognitive sophistication. The typical capuchin monkey is stubborn and determined to succeed at any foraging challenge, no matter how much time it takes or how much pain is involved. Just as a capuchin refuses to acknowledge absolute defeat in a fight and continues to threaten its victorious opponent from a distance when the fight is over, as if already planning round two, a capuchin will not concede defeat to the mechanical defenses of a plant. I have seen a young capuchin monkey manipulate the same panama fruit for more than forty-five minutes before extracting the first bit of food from it, even while getting his hands covered with the irritating hairs that serve as the plant's defense against premature dispersal. Monkeys quite often get lost from their group because they are so intent on solving a foraging problem that they do not notice the group has vanished. When confronted with a fruit they cannot open, capuchins will determinedly and enthusiastically runs through their "bag of tricks," often without pausing first to consider which motor movements would be effective in light of the specific physical properties of the particular food item. "When in doubt, pound and scrub" seems to be the motto of the foraging capuchin when encountering a food for the first time.

Perhaps the most entertaining aspect of capuchins' interactions with plants is their habit of rubbing particular plant parts into their fur. I will never forget the first time I saw this happen, in my 1990 field season. Alpha male Curmudgeon wandered into a *siempreverde* shrub (*Jacquinia pungens*) and opened the husk of one of the little green fruits with a popping sound. Then, instead of eating it he began to rub it all over his body. Soon more and more monkeys were piling into the tree. Some popped open their own fruits, and others just rubbed against monkeys who were already anointing themselves with the fruit. As they rubbed,

they closed their eyes halfway, furrowed their brows, and assumed an ec-
static expression. The group became one big ball of writhing monkeys in
a state of frenzied motion. Occasionally a monkey would become so absorbed in trying to maximize his body contact with groupmates that he
would forget to hold on to the tree and would start to fall before regain-
ing his poise and working his way back into the squirming ball of monkeys. I laughed till I was doubled over, and I was completely perplexed as
to the explanation for this phenomenon. Were the monkeys having an
orgy? Did these little *Jacquinia* fruits have psychoactive properties?

I had never seen the monkeys this excited about anything. It was remarkable that such contentious animals could tolerate one another's
close proximity; they must really be keen on this plant substance, I
thought, to put themselves so close to the jaws of their overexcited companions. Every few minutes or so someone squawked when he was unex-
pectedly slapped or nipped, and a fight would break out, but the monkeys were so anxious to resume rubbing their fur with the fruit that the
fights did not escalate or persist for more than a few seconds. Also, the
monkeys frequently became so excited that they sexually mounted and
thrust on whomever they were in contact with at the moment, then resumed their fur rubbing.

Another peculiarity of the fur rubbing was that, although there were
plenty of fruits in the tree for the taking, the most desirable fruits were
apparently the ones already in use. Occasionally a monkey would drop a
slippery, drool-coated fruit, and multiple monkeys would stare down in
shock and horror at the lost fruit. Then several monkeys would climb all
the way down to the ground to retrieve it, rather than taking the much
shorter route to pick a fresh one. An additional intriguing aspect of the
event was the way that some group members (particularly Guapo) would
take it upon themselves to thoroughly rub the substance into the fur of
infants who were too young to anoint themselves.

As soon as I got home that day, I looked up *Jacquinia pungens* and discovered that the green fruits are loaded with saponins, plant compounds
that can be used to kill fish and insects. Was it possible that the monkeys
were using these plant substances as insect repellents or for medicinal
purposes? I wrote to Linda Fedigan, the founder of the Santa Rosa capuchin project, to ask whether she had seen this behavior. She had—she
said that the Santa Rosa researchers had jestingly given this behavior

the term "group grope." The label seemed so highly appropriate for this quirky phenomenon that we started using it at Lomas as well. But why was fur rubbing such an emotionally charged social behavior? Presumably the monkeys could coat a larger portion of their body by rubbing against other monkeys, since they could not reach all parts of their body with their own hands. However, it was still a mystery why they needed to grope with so many monkeys at once, rather than with just one or two trusted partners.

That first season I witnessed several more instances of "groping" with *Jacquinia* fruits before the monkeys stopped using them for that purpose. A few months elapsed, and then at the beginning of the dry season, when some of the green fruits were starting to take on a yellowish tinge, the monkeys began eating them with gusto. They never groped with them in the same period in which they ate them. It seemed that the chemicals that had made them desirable for groping were no longer present in the ripe fruits. Perhaps the saponins, the secondary compounds that the plant had presumably evolved in order to protect the immature fruits from predation, were what the monkeys had been after when the fruits were green.

*Jacquinia* fruit is not the only plant product that the monkeys rub into their fur. During my first two years in the field I saw the monkeys use a total of eleven different substances, and in 2003 we were still adding new items to the list. Two trees, the guapinol (*Hymenaea courbaril*) and the cenizaro (*Pithecellobium saman*), produce sap that is particularly popular. Both types of trees seem prone to "open wounds"—holes in the trees into which sap and resin leak year round. When it rains, these holes fill with water, creating a diluted sap mixture that the monkeys enthusiastically rub into their fur. The fluid from both trees is black, and the monkeys completely coat their bodies and white faces with this substance, to the extent that we can no longer tell the individuals apart. Interestingly, I have seen the piglike peccaries use guapinol sap in the same way, with equal enthusiasm. It is fairly common for guapinol trees to have a hole at the base of the trunk that can temporarily house one or two peccaries or up to twenty-five squirming monkeys.

Several other plant—and insect—species are used for "group groping." Two species of the genus *Piper* (the same genus as the common pepper plant) are commonly used. The monkeys chew and crush the sweet leaves of the anise-flavored *Piper marginatum*, and they also chew the green, white pepper–flavored fruits of *Piper arboreum*. They never swallow the leaves or fruits of *Piper* plants—it seems that they masticate them just to release the relevant compounds so that they can more effectively be spread on their bodies. Perhaps the most amusing group gropes to watch are those in which the monkeys use *Capsicum* peppers. These tiny peppers are exceedingly hot, and the monkeys look as though they are in some sort of exquisite pain as they drool and grimace, their eyes tearing and their noses running in response to the sharp pain in their mouths. They drool copiously, shampoo themselves in this spicy spit, and then they go right back for more. Like the *Jacquinia* fruits, *Trichilia* fruits and *Eugenia salamanensis* fruits are chewed and "groped" with when they are green but eaten when they are mature. Citrus fruits, found on or near farms, are used for rubbing at any stage of maturation and are never eaten. Some insect species (stinkbugs, millipedes, and a few caterpillar species) are occasionally groped with as well, though insects are not used as frequently as fruits, saps, and leaves from the plant species mentioned above. In surveying the literature and talking to plant chemists, I found that there is no single compound that all of the monkeys' fur-rubbing plants have in common. But there are particular compounds that turn up with striking regularity: saponins in particular, and also flavenoids, tannins, and other complex phenolics that are reputedly effective as insect repellents.

Intrigued by the possibility that these plants contain natural insect repellents, we decided to perform a controlled experiment on ourselves, just out of personal curiosity. Every morning for a month, we bathed one leg (always the same leg) with the black, stinky fluid collected from the guapinol tree. At the end of the day, we would use a permanent marker to circle and date the new mosquito, chigger, and tick bites on the treated leg and the control leg, so that we could be sure not to double count bites received on different days. For the first week of treatment, there was a trend toward a reduced number of bites on the treated leg, but the difference was not quite statistically significant. After that, the potency of the sap we had collected seemed to diminish; there was no statistical effect detected in the remaining three weeks. Granted, we could not perform a very stringent test during this experiment because we often had to wade through rivers to keep up with the monkeys, thereby washing off the guapinol sap quite early in the day. We may not have allowed the guapinol sap to exhibit its full potency. After a month of humiliating ourselves around Bagaces with our hideous legs (by the end completely covered in Sharpie pen circles and dates that called attention to the hundreds of oozing chigger bites that normally decorated us in the rainy season), we did not have the motivation to continue with further experiments using other plant compounds, or other body parts that we could more effectively keep dry. It was just too exhausting to keep explaining our eccentric behavior to the locals, and too hot to keep our legs covered up in blue jeans.

My initial attempts to persuade plant chemists to collaborate in a study of the insect-repellent qualities of these substances failed, but fortunately other researchers have become interested in the topic. Mary Baker, who worked at Curú Biological Reserve in Costa Rica, witnessed fur rubbing primarily using citrus fruits and the leaves of *Clematis* vines.[7] She did a literature review investigating the plant chemicals present in these substances and concluded that the plants used for fur rubbing contain compounds with insect-repellent qualities. She also began a program of experiments aimed at determining what features of certain plants make them attractive to monkeys for fur rubbing. Meanwhile, Ximena Valderrama, a researcher working on *Cebus olivaceus*, discovered that the millipedes those capuchins rub into their fur contain potent insect-repellent compounds.[8] And my Brazilian colleagues have discovered that their brown capuchins rub their fur with carpenter ants to repel ticks.[9] I have recently struck up new collaborations with phytochemists at Universidad Nacional in Costa Rica, to investigate the chemistry of some of the plants used only by the Lomas monkeys.

There are a few aspects of group groping that especially intrigue me because of my interest in infant development and social learning. How do monkeys decide that a particular item is food or "medicine"? Do they use physical cues such as smell or taste, or do they learn from observation (in which case you might expect to find different medical plant-use traditions in different populations of monkeys)? The fact that monkeys everywhere choose the same kinds of plants, without much deliberation or experimentation, seems to point to nonsocial learning based on characteristics such as smell. For example, there are numerous reports from zoos of capuchins spontaneously rubbing their fur with onions, tobacco, or cranberries the first time they are presented with such items. Once when I was visiting a friend who runs a small zoo and rehabilitation center for Costa Rican wildlife, I asked her whether an infant capuchin in her care had ever had been exposed to *Piper* fruit. She said that he had not, and I asked if I could offer him some. Without a second's deliberation the monkey started rubbing his fur with the green fruit, just as the Lomas monkeys do. Never having witnessed fur rubbing herself, my friend was amazed and delighted to see this extraordinary behavior.

Of course just because the monkeys use nonsocial cues in selecting fur-rubbing substances doesn't mean that they completely ignore social cues. Further and more systematic investigation of the criteria monkeys use to select their fur-rubbing substances would be highly worthwhile. And although the insect-repellent hypothesis has by far the best fit with the data at this point, there is still much work to be done to demonstrate that the use of these substances does in fact reduce the monkeys' parasite load. Another puzzle is the issue of why fur rubbing is commonly observed in *C. olivaceus*, *C. capucinus*, and *C. albifrons*, but is rarely reported for wild *C. apella* monkeys.

Fortunately it is easy to persuade biologists to work on such fun topics. Currently Mary Baker is following up on her experiments regarding selection of fur-rubbing substances, and other researchers are exploring the social dynamics of group gropes with the aim of discovering whether there may be social functions to this behavior as well as the presumed insect-repellent function. My Costa Rican collaborators are enthusiastic about analyzing the many plants used by the capuchins. Perhaps they will discover natural chemicals that can be useful for humans as well as monkeys; I hope that if they do, the monkeys will get some credit for the discovery. Perhaps if more of us learn to view monkeys as the clever naturalists that they are, we will try harder to conserve their habitats so we can continue to benefit from their knowledge.

# CHAPTER 4

# Predators, Prey, and Personality

As the previous chapter demonstrated, capuchins are extraordinarily feisty and bold in their foraging tactics. They forage on a wide variety of plants and animals, including many prey items that require a great deal of risk, pain tolerance, and persistence to obtain. This general attitude toward problem solving (boldness, tenacity, and pugnacity) seems to extend into several other aspects of capuchin life, including their dealings with other monkeys and their dealings with animals that are not prey. Behavioral ecologists build their hypotheses on the assumption that animals "assess" (not necessarily consciously, of course) the potential reproductive costs and benefits of each of their actions in deciding what to do.[1] In real life situations, however—especially in interacting with prey and predators, when decisions have to be made with lightning speed—it may be impossible to collect accurate information on these costs and benefits and act on that information in a sensible manner. Animals' mental machinery, particularly their emotions and motivations, may not be flexible enough to ensure that an individual acts in its best interest in all contexts.

Behavioral ecologists such as Andy Sih and his colleagues have developed the view that behavioral syndromes—general tendencies to react similarly across various situations, like behavioral "rules of thumb"—play a major role in the evolution of behavior.[2] They argue that natural selection favors temperaments and behavioral patterns that enhance reproductive success in general but do not necessarily serve the animal well in all possible situations. This chapter presents a few scenes from our observations of capuchin life that typify their means of interacting with the other animals residing in the forest, including animals that could qualify as prey, as predators, and as species that seem to be ecologically neutral with respect to the capuchins.

April 13, 2001    Hannah Gilkenson, who came to work at Lomas just four months ago, is helping me do an all-day follow of Mezcla, the alpha female of Rambo's group. Mezcla is strolling down the banks of the Río Cabuyo, looking for water spiders. I hear a familiar "pok-pok-pok-pok" sound and look up to see one of my favorite sights: a basilisk lizard racing bipedally across the river, skimming at top speed along the surface of the water. These lizards are known as Jesus Christ lizards because of their amazing ability to walk (run, rather) on water. They are definitely the prettiest lizards at Lomas as well, with their shiny, almost iridescent green skin and frilly crests that give them a dragonlike appearance. Mezcla veers slightly away from the shore and into a patch of tall grass, knee-high for me and therefore high enough to hide Mezcla completely. She pounces, and captures a juvenile basilisk lizard. It is bright green, about the length of her forearm, and has toes about two inches long. Hannah and I watch in disgust as this beautiful animal is methodically devoured, bones and all, starting with the head and finishing with the tail. We cringe whenever she takes a bite, because the sound of lizard flesh and skin being torn apart makes a loud, nasty noise like fabric being ripped, quite unlike the sound of mammals or birds being consumed. It doesn't seem to bother Mezcla in the slightest. "Thank goodness that is over," I say to Hannah, as Mezcla eats the last bite. But as soon as she finishes eating that lizard, she pounces on a second. This one puts up a bigger fight, perhaps because she starts eating the tail first rather than the head. "Ugh," says Hannah, as the lizard's tail thrashes about wildly in Mezcla's hand, even after being disconnected from the rest of the body.

Other monkeys have heard the sound of lizard consumption and are flocking to the patch of tall grass to try their luck. Una and Chupacabra get lizards, and Una screams at Duende when he tries to steal hers. A lizard makes a dash for safety and dives into the water, swimming upstream well underneath the surface rather than engaging in its usual tactic of running straight across the surface to the other shore. Despite the fact that Una already has one basilisk in her mouth, she cannot resist the temptation to pursue this lizard, and she runs along the bank, following the swimming lizard's progress. She leaps into the water but the lizard sees her and swims deeper, managing to escape. Duende dives after another lizard that is also trying to escape. Fortunately for the lizards, the monkeys seem reluctant to enter water that is much deeper than their heads. Meanwhile, Chupacabra finishes her lizard and catches two more—one in each hand. Mezcla is also unsatisfied with a two-lizard lunch and is standing up, scanning for more lizards that might be hidden in the bright green grass. All of the group's juvenile monkeys are now bouncing bipedally through the grass in search of lizards, stopping occasionally to gaze longingly at the lizards held by their lucky companions. "It's funny that no one is food calling, even though this is spectacular food! Remind me to tell Julie," I say to Hannah. Our friend Julie Gros-Louis is doing research on the function of the food call (see Chapter 5).

The basilisk hunt is the sort of situation in which cooperation between hunters might have yielded a higher return. Some monkeys could have flushed prey out of the grass while others stood at the shore to intercept the escapees as they dashed for the river. However, among the capuchins at Lomas, it seems that every monkey is out for himself during hunts (though Lisa Rose has seen a few coati hunts at Santa Rosa that might qualify as cooperative).[3] I have seen no evidence of coordination: just a blur of chaos, greed, and excitement. Some studies of captive brown capuchins (*Cebus apella*) have demonstrated cooperative food acquisition (as in experiments in which two monkeys must simultaneously pull on a bar to bring a tray of food within reach).[4] But although I often see cooperation in other contexts (such as coalitionary attacks on other monkeys), I have never seen a group hunt that looked at all organized.

Perhaps in the excitement of the chase it is harder for capuchins to put together a plan for hunting cooperatively.

Chimpanzees are the only primate species aside from humans in which cooperative hunting has been documented. In the Taï Forest population of chimpanzees in Ivory Coast, different individuals perform complementary roles, with some individuals performing roles that require an ability to predict the prey's actions. The more sophisticated roles are performed only by older, more experienced individuals. Christophe and Hedwige Boesch distinguish between different levels of cooperation: "similarity" involves performance of similar actions to accomplish a goal, but with no particular coordination; "synchrony" involves temporal coordination of actions; "coordination" involves both spatial and temporal coordination; and "collaboration" involves the performance of different but complementary roles.[5] Collaboration is hypothesized to require the highest level of cognitive sophistication, since it requires some understanding of the other actor's perspective. The argument for why no primates other than chimpanzees and humans hunt cooperatively tends to rest on the assumption that they lack the cognitive sophistication for collaboration. However, since capuchins show the capacity to adopt distinct collaborative roles in bond-testing rituals (see Chapter 11), this argument probably cannot explain the lack of cooperation in capuchins. For the more common prey types (for example, squirrels and coatis), it may be that capuchins are successful enough as individual hunters that hunting cooperatively is unlikely to increase their returns dramatically. Basilisk hunts happen only about once a year, perhaps not often enough to make it worthwhile for capuchins to develop a system of collaboration in hunting that prey.

Though I haven't seen much evidence for cooperative hunting, capuchins do cooperate in defending against predators. However these incidents, like hunting raids, occur so quickly that it is often impossible to note down all of the communicative signals between the monkeys that would give us insight into whether such defensive attacks are deliberately coordinated, in the strictest sense of the word. Capuchins routinely mob predators, and during most mobbings the participants solicit one another's aid, unlike in group hunts. The most spectacular joint attack on a predator ever witnessed at Lomas was seen by my field assistants Gayle Dower and Eva Wikberg while I was away at a conference.

The full details are published elsewhere,[6] but the following is part of the jubilant e-mail I received from Gayle at the time:

> Other big news is that Mooch got attacked by a huge (2 metre) boa to-day. Don't worry she's just fine, as far as we can tell, we're not even sure if she got bitten as I think she stepped right into it. It was incredible, Susan, the monkeys moved into action so quickly, it was all over before I could even get my dictaphone out! She was there being squeezed in layers of boa and Mezcla just hurls herself at it, and yanks it's tail off and pulls and pulls and bites and bites, while Pablo hits it and yells at it and we think bites it, and a juvenile (sadly we didn't see who) helps pull Mooch from the middle . . . all the other monkeys came over too, but by the time they got there Mooch was free. Whatever my opinions of Mezcla previously she has my respect forever, she was awesome!! A truly excellent mum (though we think she might have done it for any youngster?) acting in that absolutely no-thought-for-self-must-save-baby manner. It's got to be one of the most amazing experiences of my time here (and therefore my life) so far . . . Just amazing!

**March 16–17, 1997**    I hear a chorus of alarm calls, and several monkeys take off running in the direction of the calls. I charge after them, but I cannot keep up, especially as they dash through a prickly bromeliad patch. Fortunately they keep calling continually, and I catch up with them a few hundred meters later. Several monkeys are surrounding a vine tangle in the top of a tree and furiously shaking branches and alarm calling at whatever is hidden within. I peer into the treetop and see, to my great excitement, that it is an ocelot, crouching bleary-eyed and confused in the tangle of vines and observing me warily. Joe comes over to where I am standing and whispers, "What kind of cat is it? It ran by me so fast I couldn't get a good look. Rico was in the lead chasing it through that big bromeliad patch."

"Wow!" I whisper back. "This is the closest I have ever seen one!" The cat is at least three times the size of a monkey, but nonetheless it looks pitifully distressed, both by our presence and by the fact that it is surrounded by several outraged monkeys. The monkeys are within three to four meters of the ocelot, lunging, threatening, and breaking branches. Rico and Guapo, two subordinate males, are the closest, but Hongo (another subordinate male) soon approaches, and Tattle, an adult female, is also extremely close to it.

About twenty-five minutes into the mobbing session, after the males are beginning to lose steam, Squint comes over to take a look at the situation. Squint goes berserk when she sees the predator and causes everyone else to display renewed outrage at the ocelot. Now the females are leading the attack while the males rest and observe quietly. Abby aids Squint in alarming, lunging, and bouncing at the cat. The alpha male Ichabod finally appears fifty minutes after the mobbing started and alarm calls from a distance. It is now almost 6:00 P.M.—monkey bedtime! And ocelot breakfast time. As dusk falls the cat starts looking more alert and hungry and less like a pitiful, frightened kitten. Perhaps we are letting our imaginations run away with us, but it seems as if the ocelot is eyeing the monkeys with a new sort of interest. It appears to gain confidence as darkness settles over the forest.

"This seems like incredibly bad planning on the monkeys' part," I say to Joe. "Where are they going to sleep? It's going to be child's play for this cat to find their sleeping site now."

Apparently the monkeys have suddenly thought of this too. They dash off toward the river as fast as they can, leaving Ichabod and Rico to monitor the cat. By 6:15 the monkeys are settled into an *Anacardium* tree for the night.

The next morning we arrive early, expecting total chaos and fearing the worst. My first focal animal is Ichabod, the alpha male. At 5:30 A.M., just as it is getting light, he starts alarm calling. While continually alarm calling and occasionally stopping to sniff branches, he dashes back to the tree where the cat was last seen the previous evening. He seems furious; I have never seen him so agitated. When he arrives at the scene of the mobbing, he emits a series of lost calls. Is he calling for male allies to help him search for the predator, or is he searching for someone who had been eaten by the cat in the night? Having confirmed that the cat is not where it was last seen, he starts to retrace in reverse the route the monkeys had taken the day before when they had chased the ocelot from the waterfall, across the bromeliad patch, and into the tree. As he searches, he emits alarm calls at a rate of fifteen times per minute and lost calls at a rate of one and a half times per minute—several hundred times the usual call rate. Finally, after an hour of frantic searching, he gives up, stops vocalizing, and returns to the group, which has hardly moved from its sleeping site. While I was following Ichabod, my field assistant Todd Bishop has done a quick census and confirmed that no one has been eaten.

Then, what started as an incredibly exciting focal follow turns into the most boring focal follow of an adult male that I have ever performed. All three of the group's males doze for a couple of hours at 10:00 A.M. and then have another long nap in the afternoon, as well as a few cat-naps, such that they spend 19 percent of the day sleeping. Capuchins almost never sleep during the day, so this is extraordinary. More unusual still is the fact that the females do not sleep: they simply go about their business. It seems that the males must have stayed up all night worrying about the ocelot while the females and juveniles slumbered, leaving the sentry duty to the males.

This incident provided further support for a point that Lisa Rose made about the benefits of males to capuchin females in her study at Santa Rosa.[7] Although capuchin males eat a lot and often have priority over females in access to preferred foods, they provide a valuable service by being extremely vigilant in detecting predators and also by valiantly defending against them. With multiple males looking out for predators, females can relax a bit and devote more time to foraging and resting than would otherwise be the case. Although males are the primary defenders against predators, females sometimes participate quite actively as well. For example, we once saw the normally mild-mannered subordinate female Wiggy rush up to a pair of tayras (large weasel-like carnivores) and grab one of them by the tail, yanking hard.

Many of the capuchins' interactions with other animals seem extremely dangerous, and it never ceases to amaze me that their mortality rates are as low as they are, given their bravado when confronting potentially harmful creatures. One topic that interests me greatly is the process by which young capuchins correctly categorize the animals they encounter. How do they decide which animals are prey, which are predators, and which are feeding competitors? How do they know when to eat, bully, or flee from a member of a particular species? Instinct is typically assumed, but as is true for many primates, capuchins are quite likely to have considerable learning to do.

January 8, 2006     Lydia Beaudrot and I are following Dante, Diablita's year-old infant, as she forages on insects in sticks. She discards her stick and glances in a tree hole to see if there is anything good to eat in there. She startles a bit, moves to stand a body's length from the hole, and starts producing snake alarms while continuing to look into the hole. She alarm calls eleven times, but the calls have a faint and tentative quality to them, as if she is not absolutely convinced that this is a snake. Her older sister Dali, now four and a half years old, rushes over to investigate. She emits a single snake alarm call and then reaches into the hole, grabs a snake, and flings it to the ground, where it lands between Lydia and me, just inches from my feet. I leap back because it seems at first glance to be a baby *cascabel*, or rattlesnake. However, on closer inspection I see that its eyes are rather bulbous, its markings are not sufficiently triangular, and it has no rattle. I get out my camera and photograph it. Anne Bjorkman, another field assistant, later e-mails the photo to a herpetologist she knows, who identifies it as a blunt-headed vine snake, *Imantodes gemmnistratus*. This is a nocturnal species that tends to hide in holes or crevices during the day. It has rear fangs and eats lizards and frogs.

Such observations, when the monkeys themselves are not entirely sure how to categorize an animal they encounter, are always interesting. In this case it appeared that both Dante and Dali had some doubts, but the more experienced Dali, after a few seconds' consideration, decided it was not a rattlesnake and demonstrated her lack of fear by grabbing the snake and dispensing with it. It is likely that young animals learn quite a bit about how to categorize the plants and animals they encounter by observing the reactions of their groupmates. When I looked at the proportion of alarm calls that were directed toward dangerous versus nondangerous animals, I found that juveniles under one and a half years of age directed 19.4 percent of their alarm calls at nondangerous animals, compared with 10.6 percent for juveniles older than one and a half and about 5 percent for adults.[8] It seems that capuchins start out by overgeneralizing and then narrow their range of creatures classified as dangerous as they acquire experience. Dorothy Cheney and Robert Seyfarth discovered a similar pattern in the development of vervet monkey alarm calls.[9]

We do not know to what extent social influence, as opposed to individual experience, is responsible for this age-related change. Certainly direct interactions with potential predators could be extremely costly if a youngster miscalculated, whereas reliance on social cues would be less costly, assuming that the models know more than the infant does. Young capuchins definitely pay attention to older individuals' reactions to animals encountered in the forest. But in the absence of any control over the monkeys' access to (and hence quality of experience with) various animal types, it is virtually impossible to measure the effect that social learning has on the acquisition of attitudes toward other species.

Experiments on captive rhesus monkeys indicate that both social learning and innate propensities play a role in the acquisition of snake phobias.[10,11] Captive-born individuals showed no fear of snakes unless they had seen a fellow rhesus monkey reacting fearfully to a snake, in which case they immediately acquired a permanent snake phobia. On the other hand, fooling the monkeys into thinking that they were seeing another monkey reacting fearfully to flowers did not produce a flower phobia; this suggests that the animals are specifically "prepared" to learn to fear their natural predators.

Snakes pose a particular categorization challenge to monkeys, since size is not a good predictor of how dangerous they are. We do not yet know what cues they use to distinguish between dangerous snakes (boas and rattlesnakes) and nondangerous snakes (practically every other snake residing in the forest, including viper mimics such as the vine snake found by Dante and Dali). Since broad nonsocial cues such as size and patterning are of little use, it seems likely that the monkeys rely on social cues to help them fine-tune their decisions about which more subtle cues to heed in identifying a snake as dangerous.

I have often felt baffled by the monkeys' interactions with another potentially dangerous category of animals—birds. The monkeys are eager to eat the eggs and fledglings of practically any avian species, but of course some raptors are capable of carrying off an infant monkey. Sometimes the monkeys will physically attack enormous nesting birds to get them off the nest and steal their eggs, and other times the monkeys are intimidated by tiny, sharp-beaked birds. Monkeys will also physically attack extremely large and dangerous birds, rather than fleeing as they would from a feline carnivore. It must be daunting for juvenile monkeys to decide how to react to the birds they encounter. Following are a few incidents we recorded that demonstrate the monkeys' interactions with birds, both as prey and as predators.

June 15, 1992     I hear a chorus of vocal threats and run toward the sound. When I get to the source, I see a throng of monkeys in a vine tangle, surrounding a large nest that looks vaguely similar to a squirrel's nest. The monkeys threaten the nest and lunge at it. Someone above the nest shakes branches, and Curmudgeon gets beneath it and shoves it. He positions himself just under the nest and begins to push it up and down violently. This action causes the enormous bird in the nest to flop up and down, like a human on a trampoline, and I realize that I am looking at a curassow, an endangered species. This bird is about four times the size of a capuchin, and it thrusts its neck out angrily toward the monkeys who are lunging toward it, effectively blocking them from gaining access to the interior of its nest. I am amazed to see its nest so high up; I had assumed that curassows nested on the ground.

Abby comes over to help Curmudgeon operate the "trampoline," and feathers really start to fly, but the bird refuses to be dislodged. Curmudgeon changes tactics and starts to dismantle the nest from below. He stays just under the center of the nest so the bird can't reach over the side and peck him. He starts removing fistfuls of nest material and feathers, and finally he pulls out a giant egg that is almost the size of his head. This spectacular prize immediately attracts a throng of beggars, whom Curmudgeon slaps out of his way. He retires to a rough-barked branch and starts to delicately scrub the tip of the egg to make a hole. Guapo, meanwhile, makes a beeline for the bottom of the nest, removes a second egg, and flees. With both eggs gone, the mother bird makes a piteous squawk and flies away. Paul Bunyan looks in the nest but comes away empty-handed. The rest of the group closes in and rips the nest to bits, hoping to find something that was missed.

Curmudgeon, no doubt heated by his exertions in playing trampoline with the curassow, takes two long, luxuriant baths in a cool stream while clutching his egg under his body. Four juveniles observe him in envy but make no attempt to steal the egg. Guapo, on the other hand, gets far less respect and must work in great haste if he wants to consume his prize. Surrounded by beggars, he loses his cool and breaks his egg, dropping half of it to the ground. Egg yolk splatters everywhere, and the rest of the

Curmudgeon slurps down the contents of a curassow egg he has just taken from beneath the mother bird. Photo: Susan Perry.

monkeys charge over as if a piñata has been opened and lick the yolk off of the vegetation. Curmudgeon succeeds in making a neat hole in the tip of his egg and raises it above his head so he can slurp the contents into his mouth. After doing this for a while, he widens the hole so that he can insert his whole face. When he has drunk his fill of this enormous egg, he drops it and it lands on the startled Kola. Immediately various other group members rush over for a share of the spoils, and the ensuing scene looks a bit like tackle football, with yolk-covered monkeys running and screaming, clobbering one another for the privilege of holding the biggest pieces of eggshell. When the last traces of yolk have been consumed, the monkeys return to the nest and utterly demolish it.

Capuchins raid all manner of birds' nests, both for eggs and for fledglings. Occasionally we have seen them eat adult birds as well, which they catch in flight or else pounce on in the nest. *Cuyeos* (nightjars), for example, nest on the ground, attempting to blend in with the leaf litter, and they are relatively easy prey for a sharp-eyed capuchin. Capuchins have also been known to eat parrots or magpie jays. But capuchins can be prey as well as predators when it comes to birds. Several species of raptors elicit antipredator behavior from capuchins, such as spectacled owls and collared forest falcons.

**February 4, 1992** I am following Kola, one of my favorite juvenile males, who is cheerfully engaged in threatening me from on top of Guapo's back. A bird alarm call jolts them out of this pleasant pastime, and they break off their coalition to rush toward the sound. They pause periodically to stare alertly in the direction from which the alarm call came, and to threaten that way. Kola jerks his head toward Guapo to solicit his aid, as we hear more alarm calls. Suddenly a loud, shrill whining call from the bird itself helps the monkeys locate it. It is an enormous raptor, and Kola, three adult males, two adult females, and five more juveniles are staring at it, horror-stricken, as it flies away. As it takes flight, the females dash toward their babies, who are simultaneously running toward their mothers, and the infants leap onto their mothers' backs as the moms crouch down to encourage them to come to safety. While the females retreat with their infants, the adult males Guapo and Paul Bunyan run after the bird, following it into the tree where it has come to roost. The males continue to threaten the bird, breaking branches and chasing it, until it flies far from the group.

On a separate occasion I saw an adult male of Pelon group physically attack an enormous falcon, biting and wrestling with it until the bird took flight. Given this kind of stupendous bravery in dealing with truly dangerous birds, it is astonishing and amusing that capuchins are so thoroughly intimidated by one particular bird species, a tiny, brown, unremarkable-looking bird that lives in the forest. Though I always ask visiting ornithologists if they can identify it, no one has ever been sure what it is other than the ubiquitous "LBJ" (little brown job), and my best guess is that it is some sort of flycatcher. For lack of a better name, we call it the "Egmont bird" because its call during mobbings is identical to a five-note sequence from the violin part in Beethoven's Egmont Overture. The bird is quite small (somewhere between a large hummingbird and a sparrow in size), but it has a long, sharp beak and it guards its nest with extraordinary valor.

Because the monkeys cannot resist an opportunity to raid a nest for eggs and fledglings, every year they clash with this bird on a regular basis around nesting time. As soon as they hear its call, their eyes widen and they spasmodically clutch the branch they are standing on, scanning frantically for the bird. If they see it coming near, they sometimes flip under the branch in terror, hoping to avoid a peck. One of the funniest and most pathetic capuchin moments I have ever witnessed was when Pablo, the venerable long-term alpha male of Rambo's group, was terrorized by an Egmont bird for nearly an hour. Pablo is normally cool and dignified, but he was driven so crazy by the bird's constant pecking and dive-bombing that he ignominiously fled. The bird chased him for about forty minutes, and he was nearly half a kilometer away from the rest of the group before it relented and gave him some peace. Pablo alarm called and flailed and swatted at the bird frantically the entire time; I have never seen a male so beside himself except when under physical attack by other monkeys. I do not understand why capuchins, who are completely unperturbed by the pain of stings during massive wasp attacks, and who are bold enough to physically attack major predators, are terrified of this tiny, harmless bird. But as much as I love capuchins, it is a refreshing change to see another species get the upper hand occasionally, especially as the Egmont bird does them no harm.

One reason I find Sih's behavioral-syndromes theory appealing is that it could explain why capuchins devote considerable time and energy to tormenting creatures that are neither potential prey nor potential predators. These actions may be a by-product of the pugnacious capuchin personality that has been selected for because it enhances hunting success and political success, and also intimidates predators. Another possibility is that such ecologically neutral animals are incorporated into displays of bravado to enhance the reputations of the individuals performing them. Or perhaps, in the case of cooperative harassment, the victims of capuchins' attacks serve as social tools that allow the monkeys to solidify their social bonds during practice alliances, which are warm-ups for future situations in which the outcome of the coalition really matters.

Many of the animals that capuchins pester clearly pose no immediate danger to the monkeys. The only forest animal that seems virtually free of harassment by capuchins is the agouti. Agoutis' alarm calls are attended to by capuchin monkeys because they are hunted by many of the same predators, and it could be for this reason that they generally do not fall prey to the monkeys' ill temper. All other vertebrates are fair game for the monkeys' curiosity and pestering, even if they are nocturnal and must be awakened by the monkeys before they can be tormented. Tropical porcupines, for example, are typically slumbering in the crook of a tree when discovered by the monkeys. The monkeys seem to recognize the danger of the quills, since they alarm call at them when they find them on the ground *sans* porcupine (that is, when the rest of the animal has long since decomposed). When pestering a sleeping porcupine, they move forward cautiously and gingerly grasp a quill between thumb and fingers to give it a hard tweak before darting back.

The monkeys also arouse anteaters from sleep to harass them. Although capuchins will devour a baby anteater when they find one alone, they never prey on adult anteaters, which are about twice the size of a monkey. Typically the dozing anteater will attempt to ignore the slaps and squeaky threats of the monkeys for a few minutes, but after enough tormenting, it will start to get irritated.

**March 28, 1993**     Guapo is resting on a branch when Paul Bunyan, the alpha male, threatens him from a distance. Guapo leaps up and headflags to Paul for assistance before turning to glare at a slumbering anteater. Guapo shakes a branch at it and threatens it repeatedly. Paul Bunyan rushes over to threaten the anteater too, and grabs it firmly by the tail, giving a hard tug. This jolts the anteater into consciousness, and he emits a horrifyingly sinister, hissing screech that causes Guapo to back hastily away from it. I cannot tell whether the anteater has actually bitten Paul Bunyan, because leaves are partially obscuring my view, but he seems extremely upset and continues to threaten the anteater vocally and to lunge at it.

Although it was Guapo's idea to attack the anteater to begin with, he now seems content to leave all the dirty work to Paul. Guapo backs farther away and continues to watch with great interest as the anteater lunges back at Paul and bristles while Paul threatens him again. The anteater begins to calm down slightly, but Paul goads it into action by lunging and threatening so that it vocalizes once more. Abby and

Squint, the two highest-ranking females, rush over to help Paul threaten the anteater. Emboldened by the encouragement of the females, Paul swaggers up to the anteater and slaps it hard in the stomach. The anteater turns and starts to flee, but Paul grabs it by the tail. The anteater recoils and tugs its tail away from Paul, but Paul once more seizes its tail when it tries to flee. This scene repeats itself for several more minutes, with Paul tugging the anteater's tail while the females threaten it loudly, until the monkeys finally tire of this activity and resume foraging.

**June 14, 2003**    Nando Campos is following Rambo's group when it finds a kinkajou mother carrying her baby. Kinkajous are smallish nocturnal carnivores, probably not a threat to monkeys so long as they are not provoked. A few monkeys attack her, knocking the baby off her back. The infant kinkajou falls to the ground, and immediately ten to fifteen monkeys, mainly juveniles, mob the infant. The mother fights back for a while but eventually flees, chased by practically the entire monkey group. Throughout the attack, which lasts about one and a half hours, the baby kinkajou makes incessant, loud, piercing squeals that seemed to enrage the monkeys. The mother makes several attempts to return and retrieve her baby but is chased away by the monkeys. The monkeys direct many alarm calls at the kinkajous.

Later, watching the video footage of the attack, it reminds me of some kind of tackle football game in which the baby kinkajou is the ball. The monkeys run in, grab it, bounce on top of it, bite it a bit, toss it around, and play tug-of-war with it. Sometimes they leave it alone for a few seconds, and then one monkey runs in and grab it and instantly several other monkeys pile on top of the kinkajou and struggle for the privilege of tossing it about. The group's adult males occasionally pass by above the fracas, glare down at the squealing baby, leap out of the tree to land on top of it, and then bound away to resume foraging. Miraculously, the baby kinkajou survives this ordeal without any obvious wounds.

Scenes such as this prolonged attack on the hapless kinkajou infant are hard for a behavioral ecologist to explain without resorting to the idea of behavioral syndromes. Certainly the monkeys could have used their time in a more productive and fitness-enhancing manner—for example, by foraging or resting. Perhaps these coalitions against harmless animals are useful for reinforcing social relationships. Alternatively, the attack against the infant kinkajou may serve to intimidate it and prevent it from aggressing against capuchins in the distant future, when it is a more formidable opponent.

A few years ago our team collaborated with Lisa Rose and several other capuchin researchers on a project investigating capuchins' interactions with all other vertebrate species in the forest.[12] What we were seeking was perhaps a cultural variation in the way in which they interact with other species, particularly with those species that are ecologically neutral. But what we found, in all social groups in each of the four study sites, was that white-faced capuchins are simply obnoxious to all animals (except for agoutis and sometimes deer) under most circumstances. It seems that it hardly even occurs to them that being amiable toward and tolerant of members of other species is an option, though there are a few reports of spider monkeys and capuchins interacting amiably on rare occasions at Santa Rosa. I should note that tufted capuchins seem to have a sweeter temperament and are more tolerant than are white-faced capuchins, both with other capuchins and with other species of vertebrates that co-reside with them.[13]

Why are white-faced capuchins so obnoxious? Several aspects of their feeding ecology and social organization may have come together to jointly select for their pugnacious temperament. For one thing, capuchins are omnivores. They are constantly on the lookout for new food sources to exploit, and they must be bold and quick-thinking to take advantage of new feeding opportunities. To obtain many of their foods, the monkeys must overcome a battery of mechanical and chemical challenges requiring much persistence, pain, and courage on the part of the forager. When hunting vertebrates, the monkeys have to be extremely aggressive and quick to conquer their prey. Likewise the monkeys need to be fast and bold to engage in cooperative defense against predators that have already attacked, such as when the boa captured Mooch. Unfortunately, their boldness and aggression carry over into situations in which the predator is sleeping and not bothering the monkeys at all. The monkeys would probably benefit by toning down their outrage, if they could, and letting sleeping dogs (or ocelots) lie. At the same time, there may be some advantage to the capuchins in having a reputation for such outrageous pugnacity: their hot temper may serve as a deterrent to predators that would otherwise quite willingly attack them.

White-faced capuchins, relative to their more peaceful cousins the tufted capuchins, tend to demonstrate a higher degree of cooperation in defense against predators and in defense against other capuchin monkeys.[14] As Chapters 7 through 9 will reveal, collaboration among males in attacking other males is a critical feature of white-faced capuchin life—one that seems to be absent for the most part among tufted capuchins. It is possible that the greater reliance of white-faced capuchins on hunting vertebrate prey is somehow intertwined with the evolution of extreme pugnacity in dealings with members of their own species, but it is hard to guess the extent to which these traits are related evolutionarily.

It is worth noting that capuchins are not always insufferably obnoxious. They can be highly tolerant of pestering by groupmates during foraging, and like most primates, they enjoy many tender moments spent in grooming, cuddling, and play. Their loyalty is extremely touching as well, particularly when an old, battle-scarred male comes to the rescue of a youngster who is scared to cross a road or a tree gap, or when a female offers milk to an infant who cannot locate his or her own mother. But the capuchin temperament definitely has an edge to it—a quickness to anger and to action, and a refusal to submit to defeat—that carries over from the foraging context to dealings with groupmates.

The capuchin temperament both shapes and is shaped by the quality of the social relationships these monkeys have with their allies, their monkey enemies, their predators, and their prey. The capuchin's hot temper and persistence make the stakes in conflicts quite high. The risky nature of the monkeys' interactions with their enemies necessitates reliance on allies, whether the conflict is with a boa constrictor or another monkey. It is the importance of alliances and cooperation that makes capuchin life so socially complex. And to trust an ally, a monkey needs to understand the ally's temperament and motivations. In the following chapters, I focus on the monkeys' dealings with fellow capuchins, starting with the nuances of their communication and proceeding to their relationship dynamics and social organization.

# CHAPTER 5

# Capuchin Communication

When I met my first capuchin monkeys back in 1990, I was bewildered by their vocalizations, which sounded like bird calls and sometimes more like electronic gadgets than like the monkey sounds I was familiar with. Observers can use their intuition to interpret, at least roughly, many of the calls made by Old World monkeys such as baboons and macaques: threats sound threatening and friendly vocalizations sound soothing. But the calls of many New World monkeys sound birdlike, and most of them cannot be faithfully reproduced by humans.

Slowly, by matching calls to their context, I began to piece together the meanings of most of the capuchins' calls, and found that some have entirely different meanings than what I would have guessed just by listening. For example, the vocal threat is a cute, high-pitched squeak accompanied by a wagging of the tongue—something that tends to inspire laughter rather than fear in human observers. And when adult females groom the alpha male in the company of their infants, they produce a loud, horrible-sounding gargling noise that is more like a rhesus monkey's threat than an affiliative vocalization. It was only after months of careful observation that I was able to train myself to identify the capuchins' calls and reliably predict what the monkeys and their interaction partners would do next after a vocal exchange. The monkeys' gestural communication is somewhat more intuitively clear, though some of their especially creative gestures, such as hand sniffing and eyeball poking, left me baffled (see Chapter 11).

What might monkeys be telling each other with their calls and gestures? Early primatologists assumed that these signals were "hardwired," involuntary, and purely emotional expressions: "I'm angry," "I'm frightened," and the like.[1] Evolution had shaped these signals, and recipients' appropriate responses to them, because an animal's emotional state is a good predictor of what it will do next: Attack if angry, flee if frightened, and so on. Two rather divergent lines of thinking about animal communication emerged in the 1980s. Sociobiologists entertained the possibility that many signals were analogous to advertisements, functioning to induce the recipient to behave in ways advantageous to the signaler and bearing only a rough relationship, if any, to the truth.[2] Meanwhile, a handful of primatologists began to suspect that some primates' calls referred not just to the caller's emotional state but also to objects in the external world.

Robert Seyfarth and Dorothy Cheney, following up on work by Thomas Struhsaker on vervet monkeys, conducted a series of ingenious playback experiments that challenged the view that different predator alarm calls in vervets reflected nothing more than different levels of fear.[3] By observing monkeys' reactions to taped vocalizations played from hidden speakers, they showed that the animals reacted appropriately to different alarm calls (for example by climbing trees in response to a leopard-elicited call) even in the absence of the predator itself or any other cue. The idea that we can know what's going on in vervets' minds when they give alarm calls is still debatable, but everyone agrees that at the level of evolutionary function, the calls are referential—they refer to something in the external world, like much of human language does, rather than being expressions of purely internal emotional states.

Two more recent developments in animal communication theory influenced Julie Gros-Louis and me as we set out to crack the capuchins' code. First, the long-accepted assumption that primate calls, unlike many birds' songs, were "innate" or "unlearned," was being increasingly questioned.[4] Evidence was accumulating to show that youngsters must learn some aspects of their communicative signals, as they must learn some of the features that identify predators. We were particularly curious about hints of differences between capuchins at separate field sites, in the forms and usage of particular calls: could these be due, at least in part, to social learning rather than genetic differences?

Second, the role of signals in social interactions was receiving increasingly sophisticated theoretical attention. Instead of assuming that these signals were either involuntary emotional outbursts or self-serving "propaganda," theorists began to develop mathematical models of the evolution of stable signaler-receiver systems. For example, Joan Silk and colleagues showed how signals of "benign intent" could evolve and persist.[5] Suppose that animals can benefit from friendly interactions, but that dominant individuals can gain an even greater benefit (in any single interaction) by feigning friendly intentions and then attacking a duped subordinate—the view proposed by the original social intelligence theory, the so-called Machiavellian, intelligence theory. What is to keep such "lies," which ultimately undermine cooperation, from becoming the norm? Assuming that the animals remember past interactions and shun liars, and that dominants benefit more from repeated friendly interactions than from an occasional successful surprise attack, evolution will produce honest signals of benign intent. These signals can be "cheap" (for example, soft vocalizations) *if* their honesty is guaranteed by the long-term value of a good reputation. In other contexts, animal signals may require a cost to guarantee their honesty, as I discuss in Chapter 11.

In this chapter I describe what we have learned at Lomas Barbudal about three classes of vocalizations: predator alarm calls, food-associated calls, and trills. In addition I describe capuchin communicative gestures associated with postconflict reconciliation, coalition formation, and other, more enigmatic forms of relationship negotiation. Finally, I discuss the possibility that capuchins also communicate using olfactory messages in urine.

In January 1999 I returned to the field, after an eighteen-month absence, to conduct a six-month study of vocal communication in collaboration with Julie Gros-Louis, my former field assistant, who was now doing her own research on capuchin communication. We had set aside this field season for Julie to do some experiments on food calls and for me to collect enough high-quality recordings for both of us to use for future playback experiments. Joe and I had planned a low-key field season for ourselves because we were bringing six-month-old Kate with us, despite warnings from many quarters that bringing an infant to the field was a recipe for disaster. Thus far parenting had been more than a little stressful, since Kate was so colicky that no day care provider would accept her, and so we had decided to try this field season without any assistants, to avoid inflicting hours of Kate's wailing and our sleepless nights on innocent bystanders. Joe and I planned to take turns caring for Kate and watching the monkeys.

We would share a house with Julie, who is also Kate's *madrina* (godmother). Julie has the patience of a saint and was not at all daunted by the prospect of living with a constantly screaming infant. The three of us quickly settled into a comfortable routine, sharing both the domestic chores and the research tasks. It is always such a pleasure to live and work with old friends. We all have a steadfast commitment to the research and to each other. There was never any bickering about who would do what when; whenever a problem arose, the first person to detect it simply accepted the challenge and cheerfully solved the it, with no resentment about whether it was his or her turn to deal with that sort of problem.

Even under these ideal living conditions, the research was more of a challenge for me this time than in any year before or after. I had engaged in no physical exercise during the previous eighteen months, nor had I slept more than five and a half hours a night during the past six months. My doctor had advised me to nurse Kate during my fieldwork, for the immunizing benefits and because she was severely allergic to both cow and goat milk. Because on alternate days I spent all of my daylight hours with the monkeys, Kate wanted to nurse most of the night, so it was rare for me to get more than a couple of hours of sleep.

Much as I love fieldwork, it was extremely demoralizing to have to crawl out of bed at 3:00 A.M., having slept as little as twenty minutes on some occasions, and leave my sleeping family to drive alone to the forest. I would stagger about for thirteen hours, just barely keeping up with the monkeys while carrying a load equaling about one-third of my body weight on my back (six liters of water, tape recorders, directional microphone, camcorder, enormous playback speaker, spare tapes and batteries, plus all my normal field gear). I was so tired some days that I could barely keep my balance crossing rivers and climbing cliffs, and I worried constantly about what I would say to the National Science Foundation and John Mitani (who had generously loaned me the speaker) if I inadvertently dropped $5,000 worth of electronic equipment in the river. Fortunately I survived the field season without a major mishap (aside from a bit of damage to my lower back). Although it certainly was not one of my most productive seasons in terms of data yield, it was a tremendous relief to know that, despite the difficulties entailed in juggling parenting and research, I would not have to give up my monkey research to have a family. I was still a fieldworker, and always would be.

First on our 1999 agenda was puzzling out the "meanings" of capuchin alarm calls. Unlike vervet monkeys, with their variety of acoustically distinct alarm calls, white-faced capuchins have a single basic alarm call. However, it is subtly modified in different situations, raising the possibility that the variations refer to different predator types. Julie and I could reliably distinguish only snake, raptor, and human alarm calls after several months in the field, and we wondered whether there were other, more fine-tuned acoustic differences between calls used in different contexts that we were simply missing. After all, we knew that Japanese macaques can detect many more categories of coos than humans can; each species has evolved to perceive and attend to different sorts of auditory inputs.[6]

As it turned out, it took me years to acquire enough high-quality calls from each context and from enough subjects to perform a rigorous acoustic analysis, and by the time I had a large enough data set to permit such an analysis, I was too busy with an appointment to the Max Planck Institute and my social learning research to do the analysis myself. I hired a postdoctoral fellow, Claudia Fichtel, to perform the acoustic analysis while I analyzed the data on behavioral responses to alarm calls. Claudia had extensive experience working on the acoustics of alarm-call systems in other primate species, so she was perfect for the job. Her work revealed that there were two broad categories of alarm calls that were readily distinguishable by a trained human ear: one for aerial predators, humans, and other capuchin monkeys, and another for terrestrial predators (primarily felines and canines) and snakes, but also, on rare occasions, for caimans and monkeys who were physically attacking the caller.[7] Within both categories there were subtle but statistically significant differences in acoustic variables, such as the pitch, duration, and amplitude of the calls, that corresponded to specific predator types.

The finding of acoustic differences alone is insufficient evidence to conclude that the signals are referential, however. It is also necessary to show that the listener perceives them differently and responds accordingly. Ideally we would execute playback experiments and record the responses to all of these call types to see whether the responses differ, but we have not yet done that. I did, however, record the response of my focal animals to each alarm call that they heard, in situations in which the predator had not yet been seen by the focal animal. This is a bit like doing a naturalistic playback experiment, though admittedly the experimenter has less control over the situation than would be the case in a playback experiment.

Monkeys responded to predators in the second category (the terrestrial predators, snakes, and so on) by mobbing them. In this category, the responses were typically not so immediate: the animal would glance toward the caller, finish what it was doing, and then move toward the predator to join in the mobbing display. Such predators generally are not dangerous once detected. It is debatable whether the alarm calls in category two are truly referential, since the acoustic differences between calls in this category do not correspond to distinctive responses by the listeners; rather, they mob in all cases.

In contrast, with predators in the first alarm-call category, immediate and quite specific responses are required because these predators are dangerous and act quickly. Monkeys instantly respond to bird alarms by looking toward the alarm caller, looking up at the sky, or both, and then sometimes by swinging under a branch if the predator is nearby. The monkeys respond to unfamiliar humans by moving away and higher up in the canopy and by shaking branches and monitoring the humans, ducking behind branches in between taking peeks. Females and infants respond to monkey alarms and to sightings of unfamiliar male capuchin monkeys by grabbing their infants and fleeing silently. The variety of distinctive responses to these predators and to the alarm calls elicited by these predators (prior to seeing evidence of the predators themselves) provides further evidence that such alarm calls are indeed referential.

The most common vocalizations that one hears when spending a day with capuchins are food calls. A troop of foraging monkeys constantly produces pleasant little peeping noises that sound to the human ear like happy commentaries on a good meal. I had initially shied away from studying these food calls because the monkeys do not move their mouths when they food call, and because they all call at once, as if conversing with the other foragers. The effect is a bit overwhelming for a fieldworker; it feels somewhat like trying to take notes on who says what at a conference of ventriloquists. But Julie, always up for a good challenge, decided to tackle the function of the food call by approaching it from different angles.[8,9] Not only did she collect naturalistic data but she also performed two sets of experiments. The first set of experiments consisted of playing back recorded food calls to the monkeys and recording their reactions. In the second set of experiments, Julie covertly placed food in the forest for monkeys to find, manipulating the types and quantities of food to determine what conditions stimulate food calling.

At first we were reluctant to perform such experiments, because we did not want to modify the normal diet of the monkeys, and we most definitely did not want the monkeys to associate us with food, because that might cause them to rob our backpacks or to accept food from tourists or, worse, from poachers trying to capture young monkeys to sell as pets. Julie used only foods that were part of the monkeys' typical diet, and she presented them in quantities that would not dramatically alter their normal diet. She was very sneaky in placing the food for them to find. They never saw her removing the goodies from her backpack: she would anticipate the direction in which the group was moving, run well ahead of them, remove the food, and then remove herself to a safe distance from which she could observe the first monkey finding the food.

Julie planned to test two alternative hypotheses for food calling. The first hypothesis was that the calls functioned generously, to alert others to the location of sharable food. In that case the playbacks of food calls should draw listening monkeys toward the hidden speaker. In addition, monkeys finding large food items should be more likely to produce food calls than monkeys finding small items of the same type. The second hypothesis invoked enlightened self-interest rather than public-spirited benevolence: food calls functioned to announce ownership, saying, in effect, "This is mine—please stay away." In this case, listeners might prefer to find their own food rather than tangle with the caller. Previous work on white-faced capuchins by Sue Boinski and Amy Campbell at two other Costa Rican field sites had produced evidence for this hypothesis. They found that food calling tended to increase the spacing between foraging individuals and also to decrease the likelihood that the caller would receive aggression from approaching individuals.[10]

The results of Julie's food-placement experiments were bad news for the generosity hypothesis. The monkeys were no more likely to call upon discovering a large amount of food than a small amount. Also, the type of audience present (categorized according to number of individuals present, dominance rank relative to the caller, age, and sex) did not affect the likelihood of calling upon discovering food, as one might expect if the generosity hypothesis were correct (for example, adult females should be more likely to call when surrounded by other females and juveniles, some of whom are their relatives, than when their only neighbor is an immigrant male). However, the food type did affect the probability that monkeys would call; monkeys called more when they discovered fruit (mangoes) than when they discovered eggs or caterpillars.

Two results were consistent with the ownership-announcement hypothesis. First, when silent individuals were approached while eating, they were more likely to start calling if approached by a high-ranking monkey than if approached by a low-ranking one. Second, calling reduced the probability that the approaching monkey would attack the discoverer of the food. In other words, it appeared that monkeys who do not announce their food discoveries run the risk of dominant monkeys aggressively attempting to steal the food when they discover subordinates gluttonously devouring a secret stash, whereas food ownership typically is respected if possession is announced in advance.

The results of the playback experiments were more consistent with the generosity hypothesis than the ownership-announcement hypothesis. The monkeys tended to look toward and approach the speaker more often when a food call was being played than when a control vocalization was being played. An independent set of experiments by Mario di Bitetti on the food calls of brown capuchins produced similar results.[11] These findings are consistent with the idea that listeners associate food calls with the presence of food, but it does not necessarily tell us what the intent of the caller was. Indeed, the apparently contradictory results of the food placement and playback experiments can be reconciled by recognizing the distinct roles of the signaler and the listener. It is possible that the signaler intends one message, "Please don't come over here, this is mine," and the receiver of the message learns, "There is food over there, but that monkey is already in possession of it." The behavioral outcome of the interaction is that the listener approaches close enough to take advantage of other food in the patch (if it is a large enough patch of food), but not close enough to interfere with the caller's foraging, out of respect for the caller's previously announced ownership. In this interpretation, the call serves both to attract the listener to food and to repel the listener from food already claimed, and both parties benefit from the call.

The other call that Julie studied was the trill, a rapid sequence of clear, high-pitched staccato notes. This call is particularly interesting because it seems to have multiple functions. Sue Boinski and Aimee Campbell had found that it functions to coordinate group movement.[12] However, much of what Julie discovered supported the hypothesis that trills are honest signals of benign intent,[13] like the rhesus macaque "girney" vocalizations that provided the empirical support for Joan Silk and colleagues' model. She found that trills were produced primarily by immature monkeys, and 75 to 90 percent of them occurred during close-range social interactions, for example when the caller was approaching another monkey before engaging in friendly behavior, such as grooming, cuddling, or nursing. Trilling by infants decreased the probability that the infant would receive aggression (mainly nips and slaps) from the monkey it approached. Trilling also increased the probability that the infant would engage in a mutual friendly interaction with the monkey it approached.

At first we were excited by these results because they suggested the possibility of socially learned, or cultural, differences in call function between Lomas Barbudal and the sites where Boinski and Campbell worked. Cross-fostering experiments, in which infant rhesus and Japanese macaques were switched at birth, had shown that some monkeys' calls could be diverted, at least slightly, into alternative functions by their social surroundings.[14] Although no one had observed the extreme experiential flexibility implied by our trill results (the same call being a group-movement coordination signal at two sites but a close-range signal of benign intent at another site), it seemed plausible.

However, as we investigated the matter further we discovered that the Lomas Barbudal capuchins also used the trill in a group-movement coordination context about 10 to 25 percent of the time. At Lomas, monkeys who were traveling and were not in close range of other monkeys would produce a louder version of the trill, typically in combination with a "burst" vocalization: a short, tonal burst of sound that seemed to modify the meaning of the trill from a signal of affiliation to a call that appeared to signify, "Let's go—follow me!" It is likely that the difference in results between the Costa Rican sites reflects not intersite differences in the monkeys' behavior so much as methodological differences in the way data were collected.

Even taking these variations into account, there may nevertheless be some genuine intersite differences in white-faced capuchin vocal usage. For example, the burst vocalization has not yet been reported at other sites, so it remains possible that this modification of the trill (by combining it with another call type) is an innovation of the Lomas monkeys. Further research would be needed to confirm this, however. Such a discovery would be exciting, because the linking of one call to another to modify its meaning would be a bit like human grammar. Currently there is very little evidence for grammar in nonhuman primate vocal communication systems. Syntax is one of the main features of human language that make it a uniquely complex form of communication.[15]

Two forms of syntax have been distinguished: lexical syntax (regular rules for arranging words, or units of meaning) and phonological syntax (regular combinations of different units of sound). Phonological syntax has been documented in tamarins[16] and titi monkeys.[17] Klaus Zuberbühler and colleagues have identified potential examples of lexical syntax in a few species of forest guenons. For example, their studies have shown that combinations of calls produce different responses than individual calls by themselves. In one study of Diana and Campbell's monkeys, responses of listeners indicated that the initial call in a sequence can act as a modifier for the subsequent call.[18] In another recent example from putty-nosed monkeys, Arnold and Zuberbühler found that calls produced individually do not carry specific meaning, as is the case for alarm calls in a number of species; rather, sequences of two distinct call types are produced in response to particular predators. Furthermore, those two call types are combined into a third sequence that is strongly associated with group movement. Therefore different combinations of two call types appear to have different meanings (that is, they are associated with different contexts and behavioral responses).[19]

For the Lomas capuchins, Julie and I have documented at least twelve types of vocal combinations, some of which are compound calls in which one call is superimposed on another, and some of which are regularly occurring sequences of discrete calls. Further analysis, and perhaps playback experiments, will be needed to determine whether the combination of calls affects the perceived "meaning" of the calls, and also whether the ordering of discrete calls affects the function.

As signals of benign intent, trills appear to be one form of communication about social emotions. What other messages about their social lives might monkeys be exchanging? As we know from our own experience as social primates, relationships have many facets. Your boss may also be your friend; your sibling has probably been both your ally and a competitor for your parents' attention. Of all the possible messages that nonhuman primates might convey about their relationships, two have received by far the most research attention. Dominance, or one animal's consistent ability to win contests against another specific individual, is often signaled by species-typical, stereotypical gestures (actually, it is more often the subordinate who ritually signals its submission). This has been known for decades.[20] And more recently scientists have become aware that group-living nonhuman primates, because of their reliance on each other for mutual protection and other services, put considerable effort into maintaining and, if necessary, repairing their friendly social relationships.

More than twenty-five years ago, Frans de Waal pioneered the study of *reconciliation* in nonhuman animals, after noticing that the chimpanzees of the Arnhem Zoo colony were more likely to approach, hug, and kiss each other after fighting than at other, otherwise comparable times.[21] Reconciliation has now been documented in almost thirty species, mostly among primates but also in goats, dolphins, and spotted hyenas.[22] Most reconciliation research has been conducted on captive animals, but enough has been done in the wild to assure skeptics that reconciliation is not an artifact of captivity.

As data documenting reconciliation in many species piled up during the 1980s and 1990s, Frans de Waal, Filippo Aureli, Marina Cords, and others developed the view that social conflict, though an inevitable result of divergent individual interests, damages gregarious animals' most valuable resources: their social relationships.[23,24] This perspective, labeled "the relational model," explains reconciliation as well as restraint during conflicts. Reconciliation repairs relationships, reassuring former opponents that they can still expect the same favors and services (tolerance at feeding sites, say, or coalitional support) from each other as they could before the fight. The relational model regards mild aggression between groupmates as a form of relationship negotiation rather than a purely antisocial force. It is clear that there is more to reconciliation than appeasement of the powerful by the weak, because reconciliations are often initiated by the dominant one in a pair of opponents, or by the animal that started the fight in the first place.

Of course, each individual's social relationships vary in value, so we would expect the likelihood of reconciliation to vary too. In macaque monkeys, kin dyads are most likely to form alliances, and also more likely to reconcile, than non-kin dyads. The most striking evidence for an effect of relationship value on reconciliatory tendency comes from an experiment by Cords and S. Thurnheer on captive long-tailed macaques, in which relationship value was manipulated artificially.[25] First, pairs of monkeys were placed in a situation in which each one could obtain a coveted food reward through its own actions. Then the apparatus was changed so that the close presence of the partner was required to get the tasty treat. Once the two monkeys needed each other, their tendency to reconcile, following conflicts provoked by the experimenter, increased dramatically. These results could be interpreted to mean that reconciling monkeys seek only short-term rewards rather than relationship repair. But in a study of naturally occurring conflict in Japanese macaques, Nicola Koyama found that for a ten-day period following each fight, grooming and approach rates between the opponents were higher, and aggression rates were lower, if the conflict was reconciled than if it was not.[26] So it seems that reconciliation, or the lack of it, can have lasting effects on primates' social relationships.

Our contribution to this question began during my original fieldwork, conducted from 1991 to 1993, when we followed at least one participant in every fight for thirty minutes immediately following the altercation, and then followed one of the participants again for thirty minutes at the same time of day on the next observation day. This protocol—the "post-conflict/matched control," or PC/MC method—was developed by de Waal and colleagues as a rigorous technique for detecting reconciliation. If animals reconcile, you would expect to see them engage in friendly interactions earlier in the PC periods, on average, than in the MC periods. And you would expect reconciled opponents to affiliate preferentially with each other rather than with their groupmates generally. During our 2001 field season, we supplemented our first PC/MC data set by doing long (ideally all-day) follows of individuals. In analyzing these data, we were able to compare interaction rates between former combatants during post-conflict periods with otherwise similar periods that had not been preceded by aggression between the two monkeys. Sure enough, when analyzing the data from the longer follows, we found a statistically significant tendency for reconciliation, especially following fights between an adult female and the alpha male (regardless of which male was alpha).[27] Though not unexpected, our findings were the first demonstration of reconciliation in the wild for a New World monkey.

One of the most noteworthy aspects of the capuchin gestural repertoire is the rich assortment of species-typical signals for coalitionary aggression. For example, the monkeys have three different postures for indicating mutual support against a common enemy. They (a) align their cheeks and shoulders side by side, (b) put an arm around the ally, or (c) stack on top of one another in the overlord position (as it was dubbed by John Oppenheimer,[28] the first primatologist to study wild capuchins). In all three situations, the allies glare at the common opponent and open their mouths, baring their teeth while emitting sporadic squeaky threats.

There are also multiple ways that capuchins solicit aid. The headflag is the most common method: the signaler quickly jerks his head toward the prospective ally and then whips it back toward the opponent. This

Curmudgeon (right) and Tattle perform a cheek-to-cheek threat directed at observers the day after Tattle's infant Eldritch is born. Photo: Susan Perry.

gesture differs from the headflag gesture in baboons, in that the capuchin does not change facial expressions when soliciting an ally: the
teeth remain bared even while glancing at the intended ally. Baboons
rapidly switch back and forth between menacing and friendly expressions, depending on whether they are facing the enemy or the ally. If the
capuchin is near the prospective ally, he or she may also pounce on the
prospective ally's back, or back into the ally to request the initiation of
the overlord position. The universal use by white-faced capuchins of
this rich assortment of coalitionary signals seems indicative of a long
evolutionary history of coalitionary aggression. A typical capuchin conflict is packed with fascinating signals (both vocal and gestural) for negotiating who is allied with whom.

January 20, 2006     As I crash through the brush, I hear screams and
run over to see what the commotion is. I see the adolescent male Simba
screaming and grimacing at Oden and Fonz, who are stacked on top of
each other and threatening Simba on the ground, a few inches from

Simba's face. This is surprising, since Fonz, the alpha male of Pelon group, virtually never takes part in disputes among his descendents. Oden is Fonz's adolescent son, and Simba is Fonz's grandson. Both Oden and Fonz bare their teeth at Simba, and Fonz bounces on top of Oden as they glower at Simba. Simba squirms agitatedly but seems to want to stay in close range, despite the fact that Oden keeps reaching out and slapping him on the head or tugging hard on his ear. Simba backs away a bit, still screaming constantly, and then Oden breaks out from under Fonz's body and lurches forward toward Simba, who rears back in a panic and screams some more. Oden grasps Simba's left nipple and palpates it, while Simba falls over backward, exposing his erect penis. Oden releases Simba's nipple and grasps his hips, causing Simba to stand up and present his hindquarters. Oden mounts him in a dorsal-ventral position, grasping Simba's ankles with his feet, and thrusts for half a minute. Meanwhile, Fonz loses interest and saunters away. When the mount ends, Oden and Simba calm down and resume foraging next to each other as if nothing has happened. Apparently I have witnessed one of those rare moments in which capuchins successfully reconcile a conflict immediately after it has occurred.

Most capuchin conflict involves such a rich repertoire of gestural and vocal signals that it is difficult to tease apart the meanings of the individual signals. This problem is augmented by the fact that many signals seem to shift in meaning according to the context in which they are produced and the developmental stage of the individuals producing them. Take, for example, the overlord posture.[28] In this posture, one monkey gets on the other's back in the same position assumed by an infant riding its mother. That is, the heads are aligned one on top of the other, facing the same direction, and the spines are also aligned, with the hands of the top monkey embracing the chest of the lower monkey. The pelvis of the top monkey is on top of the lower monkey's hindquarters, unlike in the sexual mount posture. I cannot help but wonder whether the similarity in posture with the mother-infant carrying position enhances the feeling of solidarity between two allies, since infants no doubt get a feeling of security from riding on the back of a mother or alloparent who protects them from danger.

Undoubtedly, the victim of a coalition sees the overlord posture in a different light: the spectacle of two heads neatly aligned and both baring their teeth toward the common enemy must be extremely disconcerting. The white shoulders and faces of the two monkeys merge into a "super-normal stimulus," so that the two look like one enormous supermonkey. On rare occasions I have seen as many as four monkeys stacked on top of one another, like a totem pole, all squeaking and baring their teeth at the common opponent. When confronted with such a menacing coalition, the victim typically squirms and grabs his own tail, frantically kneading and licking the prehensile part of the tail with both hands as if it is a security blanket, while anxiously watching the actions of the opponents.

There is no formal gesture of submission in capuchins as there is in social carnivores or many other primates. Although capuchins do have a grimace reminiscent of the fear grimace in rhesus monkeys (which is a formal signal of submission), the capuchin grimace grades into the bared-tooth threat display and is used even by monkeys who are continuing to defy an opponent. The capuchin temperament is such that a monkey will stoutly refuse to admit defeat except under the most extraordinary of circumstances. When a capuchin monkey flees from his attackers, he typically runs just a short distance and then spins about to bounce at and threaten his attackers again, as if determined not to let his opponents have the final word.

The nipple-palpation behavior exhibited by Oden toward Simba has its roots in mother-infant interactions, just as the overlord apparently does. Normally when an infant nurses from its mother, one of the infant's hands is busily kneading the nipple that is not being suckled, thereby stimulating milk production. The infant will switch from nipple to nipple several times during a nursing bout, kneading the other nipple all the while, until the mother's breasts run dry. During weaning, infants often respond to their mother's rejections by thrashing their tails, screaming, and reaching out to palpate the desired nipple, as if handling the nipple creates some of the feelings of comfort produced by actually nursing. Generally these nipple palpations during weaning tantrums prompt the mother to groom the infant.

Capuchin infants are suckled by females other than their mothers,[29] and they even suck and manipulate the nipples of adult males and juveniles, apparently receiving comfort from these interactions despite the lack of milk. This nipple palpation behavior extends into adulthood, and it is common to see monkeys (particularly adult females) mutually stimulate each other's nipples during the aftermath of a fight, while grimacing and producing vocalizations that sound like a hybrid of a scream and the affiliative trill vocalization. This behavior sometimes, but not always, aids the animals in calming down and engaging in a sexual interaction or play instead of further aggression.

Although genital erections such as Simba's cannot, perhaps, be termed gestures, they are interesting as indicators of emotional arousal. To an anatomist, one of the most striking aspects of capuchins is the extraordinary size of the clitoris, which has a bone in it analogous to the penis bone. Strangely, the capuchin clitoris starts out large (at birth the female's clitoris is as large as the male's penis) and atrophies over the course of development until it is barely perceptible in adult monkeys when observed from a distance.[30] Its size at birth is an incredible nuisance for researchers trying to determine the sex of infant capuchins, because the genitals of newborn males and females are essentially indistinguishable; it is only with age that the tips of these organs differentiate to the point that sexing the monkeys becomes easy.

Primatologists interested in the evolution of genitalia frequently ask me what on earth capuchins are doing with this immense clitoris. I still do not have a clear answer. Males and females both have frequent erections, not only in the context of sexual play but also when they are in stimulating social situations. For example, the mere presence of the alpha male one to two meters away will cause the penis or clitoris of every youngster in the vicinity to become completely rigid. The infants, every hair on their body erect, will start to follow the male, squirming and contorting their bodies in excitement and making gargling sounds. And erections are not limited to social excitement: often exciting food, such as a large egg, is sufficient to cause an impressive erection as well.

Although the penis and clitoris seem clearly to signal physiological arousal, these organs are not routinely used in social interaction, and they do not generally become the focus of other animals' interest. On a couple of occasions I have witnessed an adult female grasp the tail of an animal mounted on her (either a riding infant or a mounted adult) and rub it rhythmically over her own clitoris. But such observations are rare and cannot be said to be part of the standard behavioral repertoire for the species.

Among the more puzzling behavior patterns I observed during my early research was the dance. There are two variations on the dance; they share the same movement patterns but are accompanied by different vocalizations. The courtship dance, which precedes and sometimes follows copulation, is accompanied (like copulation itself) by particular grunts and squeaks. The wheeze dance, though it sometimes also includes grunts and squeaks, is distinctively accompanied by a vocalization we call the wheeze, a loud, high-pitched, whistle-like call that has a hysterical quality suggesting that the caller's voice is about to break. Wheeze dances sometimes end in pseudo-copulatory behavior (mounting and thrusting) between same-sex individuals.

The following dance between adult male Tranquilo and adult female Celeste is typical for a courting pair: Tranquilo purses his lips and walks stiffly forward a few steps on his branch, his eyes glued on Celeste. His hair is standing on end. After moving forward, he swings his shoulders back so that half of his body turns away from her while he maintains eye contact. He backs away from her and begins to grunt rhythmically while swinging his torso back and forth as if he is undecided whether to move forward or to retreat. He runs back and forth, swinging his body about dramatically each time he changes direction and never taking his eyes off of the female. Meanwhile, Celeste begins to pace back and forth on the branch as well, moving forward as Tranquilo moves back, and then reversing direction when he moves toward her. Like Tranquilo, she swings her body about wildly and fixes her gaze on him whenever her hindquarters reverse direction. Celeste utters a long series of breathy squeaks that ascend in pitch, while Tranquilo continues to grunt. Both monkeys have their lips pursed in what I call a duck face, so that their protruded lips resemble the bill of a duck. As Tranquilo becomes more excited, he runs more quickly up and down the branch, and his "pirouettes"—the swinging around of his body when he changes direction— become more spectacular. Celeste approaches to within five body lengths of Tranquilo, and he backs away hastily. Celeste sits down and squeaks at Tranquilo, and he zooms past her at close range, looking over his shoulder to see if she will follow.

**Rumor makes a duck face at Fonz.** Photo: Hannah Gilkenson.

Courtship dances do not always end in copulation. Sometimes an overexcited male grabs the female's hindquarters too brusquely, causing her to panic and flee. Fonz, the alpha male of Pelon group, was an expert at finessing females into the correct position without causing them to bolt. Duende of Abby's group, in contrast, would grab females by the tail and try to forcibly pull them into mount position, as if reeling in a large fish on a line. The latter technique was typically not so greatly appreciated, particularly by females who were not expecting it. I have the impression from watching these dances (which can last several minutes, even half an hour) that timing and coordination are crucial. Males who are too direct and brief in their courtship are often rejected; it seems important to dance for quite some time, and for both partners to be rather coy. Often a female will present her hindquarters to the male only to withdraw again when he moves forward too quickly to accept the offer, and more dancing will ensue. One partner must approach while the other retreats, slowly narrowing the gap until it is time to make physical contact; it was this coordinated taking of turns and exaggerated spinning that suggested the term *dance*, because it reminded me so much of a hybrid between a Victorian ballroom scene and a tango.

A noteworthy feature of the capuchin courtship display is the unbroken eye contact. During close-range affiliative interactions, capuchins do not make direct eye contact: they focus their gaze on the partner's peripheral body parts, such as hands, feet, and tail. It is only in aggressive interactions and dances that eye contact is made. During dances, both partners are highly emotionally aroused, and sometimes the male will suddenly go berserk, as if he can no longer stand the suspense, and break out of the dance to aggressively chase the female while growling at her. Even after such outbursts, however, the female may be willing to resume dancing.

Seemingly senseless displays are common products of sexual selection, the evolutionary process that produces traits that lead to success in acquiring mates, even at the expense of decreased survival prospects. The long-tailed manakin, one of the many beautiful birds that share Lomas Barbudal with the capuchins, performs a courtship routine that even the most seasoned human observer finds hilarious. As a female watches, two males flip and cartwheel over and around each other while singing a carefully synchronized duet. Only the dominant of the two males has any hope of mating if the female is sufficiently impressed; the subordinate is an "apprentice" who must spend up to several years learning this display well enough to become a "master."[31]

Most biologists believe that costly courtship displays, both behavioral and anatomical (such as bright plumage in birds) have evolved to advertise the individual's high quality to prospective mates. However, it isn't always immediately clear just what quality is being displayed. Capuchin courtship dances are loud, conspicuous, energy-consuming, and potentially dangerous, since the dancers have their eyes fixed on each other and therefore cannot look where they are going. What is the signaling value of these dances, and why are they designed the way they are? Does the degree of coordination between the partners tell them anything about their commitment to each other, or about their ability to cooperate? Perhaps the noisiness and length of the dance do tell the partners about their commitment, since they help advertise the activity to third parties who might have an interest in breaking up the interaction before it terminates in sex.

Wheeze dances posed an even more difficult puzzle than courtship dances. I have drawn on two sources of insight in trying to understand their function. First, in mammals generally, sexual behaviors are frequently "borrowed" from their original functional context—that of procreation—and used in other forms of social interaction. Second, primatologists have documented a wide range of behavior patterns, generically known as greetings, that appear related to relationship negotiation, though not necessarily to recent conflict between the animals involved. There is no widely accepted, precise definition for greetings, but generally they are stereotyped behaviors that occur when two individuals first encounter each other after a separation of undefined duration, and they often occur in isolation from other types of activities. Many of these rituals are conspicuous and bizarre. For example, baboon males frequently cross great distances to greet each other by presenting their hindquarters or handling each other's genitals, and then they part company without conferring any other sorts of social favors or costs on each other.[32,33] This decoupling of greetings from any obvious consequences makes their meaning difficult to interpret. Wheeze dances are the most dramatic white-faced capuchin greeting; others (gargles and mock coalitions) will be described in later chapters.

My early analyses of these behaviors showed that some dyads were more prone than others to perform them, but it was still unclear to me what effect the behaviors had on the animals' subsequent interactions— that is, how these interactions helped to determine the future course of their relationships. During my original fieldwork, I had often noticed that pairs of animals engaged in wheeze dances not only after severe fights but also after long separations (for example, when an adult male had been away from the group for several days during the upheaval following a dominance rank reversal).

In 1996 Joe and I were invited to a symposium on the similarities between capuchins, on the one hand, and the two chimpanzee species (chimps and bonobos), on the other. We teamed up with Amy Parish, a friend from our Michigan days, to compare nonprocreative sexual be-havior in bonobos and white-faced capuchins.[34] Amy had just com-pleted her dissertation, based on observations of the San Diego Zoo bonobos; her work had become quite well known and had helped earn bonobos the nickname "Kama Sutra apes." Female bonobos vigorously rub their sexual swellings together, apparently to reduce tension during feeding competition and other forms of conflict. They also use such in-teractions to help forge social bonds. Male bonobos engage in "penis fencing," mounting, and rump rubbing.

In the Lomas capuchins, the vast majority of sex that we observed was also clearly nonconceptive, involving same-sex pairs, adult-juvenile pairs, or females that were already pregnant or still nursing small infants. (Eventually, genetic analysis of paternity would help us sort out the mat-ing system, but our inability to pinpoint the timing of ovulation still hampers our ability to collect detailed data on female mate choice.) Our research showed that male-male dyads engage in sex far more often than male-female dyads do, and the majority of male-female sex (about 80 percent, for females who have already had two or more infants) involves pregnant or lactating females. Male-male social relationships tend to be fraught with more tension than female-female dyads. Whereas females are all related to one another, males in a group are often unrelated, or at least less closely related than the females are. Thus it is not so surprising that male-male dyads engage in nonconceptive sex four to six times as often as female-female dyads. We also found that almost 80 percent of male-male mounts occur in socially tense situations, compared with only 20 percent of female-female mounts.

Like bonobos and many other primates, capuchins have discovered that the positive physical sensations and emotions associated with sex-ual activity can be used to dispel tension in other contexts. A suite of behaviors, physiological processes, and emotions that evolved for the purpose of encouraging procreation has now acquired a secondary func-tion in promoting the formation of social relationships that are useful for gaining access to resources such as food and mates. The sexual mount position and the accompanying gestures, facial expressions, and vocal-izations are essentially identical, regardless of whether the sex is poten-tially conceptive or definitely nonconceptive (as in the case of male-male mounting). The one difference is that females virtually never mount males (I have seen this only a couple of times in fifteen years of observation), and there is reciprocity in mounting when the "copulating" partners are two males.

The reciprocity in male-male mounting between capuchins is reminiscent of what happens in savanna baboon greetings. Barbara Smuts and John Watanabe documented a phenomenon called "diddling," in which male anubis baboons handle the genitalia of other males and mount them.[33] They found that male-male dyads that exhibited high degrees of symmetry and reciprocity in the direction of mounts and diddlings tended to have more high-functioning, cooperative relationships. Smuts and Watanabe speculated that the risky nature of these interactions made this type of signal ideal for demonstrating trust, and commitment to a cooperative relationship. Capuchins rarely handle one another's genitals directly, but they do seem concerned with reciprocity in mounting, just as baboons are. When two males have a tense relationship and one mounts the other, the one being mounted appears quite nervous: the corners of his mouth twitch spasmodically, and he casts nervous glances back toward the male on his back to see what he is doing. It is no wonder that having another adult male on one's back is nerve-racking: the mounter's long canines are just centimeters from the mountee's neck, and should he suffer a change of heart and decide to attack, the mountee would be doomed.

Capuchins have mercurial temperaments, and commentary from third parties (for example, intense vocal threats by the alpha male toward an affiliative interaction between subordinate males) often causes one male to betray another. In the absence of interference from third parties, though, capuchin males often take turns mounting each other, switching positions one to three times before they terminate the sexual activity and engage in more relaxed behavior, such as grooming. Although capuchins can be exceedingly inventive regarding their gestural communication (see Chapter 11), the use of dances and nonconceptive sex to negotiate social relationships seems to be universally practiced and understood among capuchin populations. The following sequence of events demonstrates the use of dancing and sex during tense interactions among males, and reciprocity in mounting.

**December 24, 1992**     The former alpha male Curmudgeon has just been overthrown by Paul Bunyan as leader of Abby's group. All morning the males have been dancing or fighting with each other or with subordinate male Guapo, and tensions have been high. Curmudgeon watches Guapo dance with Paul Bunyan, and the two males begin to wheeze together. Then Curmudgeon emits a unique vocalization that we have recently termed "Curmudgeon's insane vocalization"—it sounds something like an alarm call, but it has a more hysterical quality to it, as if his voice is cracking under emotional strain. He fixes his gaze on Guapo, pacing toward him deliberately. Now it is Paul Bunyan's turn to be outraged, and he emits a sequence of loud staccato vocal threats in the direction of Curmudgeon.

As Curmudgeon dances his way into proximity with Guapo, Guapo leaps over Curmudgeon, who grasps Guapo's hips and mounts him. Guapo starts walking forward, and Curmudgeon thrusts rapidly while Guapo carries him away. Both Guapo and Curmudgeon purse their lips and emit sequences of squeaks that rise in pitch. Curmudgeon pauses to threaten toward Paul Bunyan, who is still glowering at the pair, and Guapo seizes this opportunity to switch roles. Now Guapo mounts and thrusts on Curmudgeon, and the two of them resume their sex squeaks. Throughout the sex, Paul Bunyan keeps up a steady stream of intense vocal threats, his whole body shaking with the force of the vocalizations.

Curmudgeon breaks contact with Guapo and starts up another dance with him. This time, instead of doing the quiet, breathy sex squeaks, they break into wheezes. Both males cast several nervous glances toward Paul, but he refrains from attacking. Guapo and Curmudgeon part company, and three juveniles in the vicinity begin mounting one another, apparently inspired by the adult males' performance. Curmudgeon grunts and turns his attention toward Paul. Once again Curmudgeon performs his special "insane" vocalization, and then he exchanges grunts with someone who is out of my view. Suddenly and simultaneously, Paul Bunyan and Curmudgeon break into a wheeze dance, both monkeys spinning and flailing their upper bodies about as they maintain eye contact. Curmudgeon begins pacing, while grunting wildly, and Paul spins and makes a duck face while watching Curmudgeon.

Further analysis has revealed more subtle differences in the functions of the vocalizations that accompany dances. For example, the sex squeak seems to function differently for heterosexual versus homosexual dyads.

In dyads involving adult males and adult females, squeaking significantly increases the probability that the dance will culminate in copulation, whereas for adult male-male dyads, squeaking during a dance significantly decreases the probability of pseudo-copulation. In same-sexed dyads, I looked at whether squeaking was being used to achieve reconciliation, but it was not—in fact, squeaking after a conflict seemed to increase the probability of renewed aggression rather than decrease it.

Whereas grunts are common in dances of any type, wheezes are rarer and are confined to social-tension situations. Not all individuals wheeze: by far the most common wheezers are adult males and high-ranking adult females. Wheezes are one of the most mysterious vocalizations in the capuchin repertoire. Approximately one-third of all wheeze duets are protested vigorously by third parties (most often alpha males), who produce intense vocal threats and run in to interfere with the dance. About one-third of those wheeze dances that are protested are then aborted, and the dancers sometimes turn on each other, apparently flustered by the arrival of the protesting party. If the purpose of the dance is relationship formation or repair, then what is the purpose of loudly announcing the dance to other individuals who might want to break it up? One possibility is that the wheeze dance is a form of bond testing that invites challenges from third parties who stand to lose something if the dancing pair succeeds in forming an alliance. In that case, the function of the dance and nonconceptive sex is to gain information about how committed the particular dyad is relative to other possible alliances each monkey could participate in. Dancing in secret would not be as informative as dancing publicly, because public dancing invites opportunities for betrayal of the dance partner in favor of another outraged alliance partner.

With so many vocal and gestural signals to attend to, it is all too easy for us to forget that we are missing out entirely on yet another mode of communication: olfaction. New World primates are far more reliant on olfactory communication than are Old World primates. Capuchins frequently wash their hands, feet, and tail tips with urine, and they are clearly very interested in chemical signals left by other monkeys, often competing physically for access to leaves soaked with another monkey's urine. It is a source of great frustration to me that this important source of information for the monkeys will always remain unfathomable to human observers. One monkey's urine smells just like another monkey's urine to me, and though I have not tried putting it to my tongue, I have to admit that I would be reluctant to lick capuchin urine no matter how high its information content.

In 2005 Joe and I collaborated with our former field assistant Fernando Campos, who had moved on to graduate school at the University of Calgary, to analyze and interpret our data on urine washing.[35] Our results clarified a few issues but also left many questions unanswered. First, males wash with urine about seven times as frequently as females do, and the alpha male urine washes far more than other males. Also, males who have risen to or fallen from the alpha position adjust their urine-washing rates accordingly, making it clear that the rank changes are responsible for the adjustment.

Males are more interested in sniffing the urine of other monkeys (particularly males) than females are. But what chemical messages are they gaining when they sniff urine? We still have no idea. You would think that males might sniff female urine to tell when they are in estrus, especially since females are fairly cryptic in their proceptive behavior and show no obvious external morphological signs of fertility, but we do not yet have good evidence of this. Males might conceivably pick up on signs of stress and physical vulnerability in their male rivals when they smell their urine, perhaps by detecting metabolites of testosterone and cortisol. Thus far, however, we have detected no tendency for males to preferentially sniff the urine of males with whom they have particularly competitive relationships. Males do obsessively sniff areas where males from other groups have recently been, which implies that they may gain information about individual identity from urine. However, there must be more information than that contained in the urine, since males are also interested in sniffing urine they have just seen deposited, by animals with whom they are intimately familiar. We need to take a more experimental approach, and also to collaborate with chemists, in order to make much further progress on the issue of chemical communication in capuchins.

Primate communication will always be one of the most fascinating aspects of primatology because of its relevance to the evolution of language and symbolic capacities generally. In addition, a proper understanding of communication is critical to understanding how social interactions create social relationships, and thus how dyadic relationships combine to create social organizations or societies. To the extent that these "building processes" are universal across social organisms, the study of communication systems in any species should be helpful in understanding human behavior. But nonhuman primates are arguably the best models for this purpose, because our own recent evolutionary divergence from other nonhuman primates means that we share much in the way of cognitive machinery and sensory capacity, and we also rely to similar degrees on hands and facial expressions to communicate gesturally. Perhaps more important still, humans and capuchins (as well as most other nonhuman primates) share many adaptive challenges: we have long-term cooperative social relationships, live in groups, have long-term parental care, engage in frequent coalitionary conflict, and operate in a social milieu characterized by a constantly shifting and complex political landscape.

Our reproductive success depends on forming dependable alliances, and on detecting subtle changes in the status of our social relationships relative to other dyads' social relationships. Such challenges call for a large tool kit for assessing partners' assets (their physical capacity to help, for example, as well as their psychological poise when confronted with challenges), their psychological compatibility and skill in coordinating behaviors, and their commitment to particular social relationships. They also call for mechanisms by which individuals can clearly indicate their future intentions and desires, so as to coordinate interactions in the immediate future, once they have determined where they stand in the group's social structure. Being such language-oriented creatures, we humans tend to overlook the importance of gesture, posture, gaze orientation, and perhaps even olfaction in negotiating our own social relationships, and we tend to focus more on the text of what people say. But, as psychologists and sociologists such as Erving Goffman[36] and Randall Collins[37] have realized, these more basic forms of social communication are likely to be just as critical to understanding the dynamics of human societies as they are to understanding nonhuman primate societies.

**February 17, 1992**     The midday siesta is drawing to a close. Nanny, the third-ranking female among the five adult females in Abby's group, is enjoying some last moments of quality social time before the monkeys rouse themselves and start the afternoon's trek through the forest to look for food. She is grooming Squint, who ranks second but spends more time grooming than any other female and is clearly the lynchpin of the female social network. Each female grooms the other for less than half a minute, then stops, turns, and "presents" a part of her body to the other for grooming. The process brings to mind human friends exchanging backrubs, except that, like most capuchin behavior, it happens at a frenetic pace, like a children's cartoon. After a few minutes it becomes apparent that Nanny is taking more turns than Squint as the recipient of the grooming, and that most of her turns are longer as well. Usually the higher-ranking female in a pair of capuchins spends more time than her companion being groomed, but recently Squint has seemed eager to please Nanny. We wonder whether a realignment of the female hierarchy is in the works. Perhaps Abby's days as alpha female are numbered, and Nanny and Squint will overthrow her and become the new central alliance. Two weeks before this interaction, when Abby had approached the grooming pair of Nanny and Squint, both had ignored her until she gave up and left the scene—a noteworthy breach of hierarchical etiquette.

Now Abby approaches the two of them and presents for grooming. Squint starts grooming Abby, then stops and presents for grooming to Nanny. In the middle of the subsequent exchange of grooming, Squint stops to sniff Nanny's hand, a bonding gesture that is currently all the rage in Abby's group (see Chapter 11). Perhaps all three top-ranking females can hang out amiably. But three minutes later, just after going back to grooming Abby, Squint abruptly stops grooming her, grabs the back of Nanny's neck, and jerks it. She then directs an open-mouth threat at Nanny, lets go of her neck, and grabs her ear. Nanny yelps, frees herself, and scampers out of the way. The Abby-Squint central alliance, it seems, is still a going concern.

Political behavior among female white-faced capuchins is usually subtle compared with the males' conspicuous bravado and occasional violence. A typical dominance interaction can consist of a subordinate's casually stepping out of the way of an approaching dominant. It took several months of observation before we had gathered enough of even such minor incidents to be confident that we knew the ranks of all five adult females of Abby's group in 1991. However, once we were tuned in to the female hierarchy and the clique structure, even the smallest social overtures or snubs, such as offers or refusals to groom, took on considerable significance.

Female capuchins apparently fret about their own and their companions' grooming decisions. Primatologists studying several species have discovered that monkeys' and apes' rates of "self-directed" behavior (scratching and self-grooming) rise far above normal levels in stressful situations, such as after fights. We found that female white-faced capuchins' self-directed behavior rates don't rise after fights, but they do rise just before and after social grooming, suggesting that grooming decisions are a source of anxiety (fights probably cause anxiety too, but a quick-moving arboreal monkey's need to grab branches with both hands and both feet in preparation for aggression or flight may suppress the urge to scratch or self-groom).[1] One interpretation of this behavior is that grooming another monkey stimulates self-grooming and scratching merely by reminding the groomer that she would like to be groomed herself. But this hypothesis doesn't explain why self-directed behavior rates also rise in females who have just *been* groomed and who are therefore very likely to be "asked" to take a turn as a groomer.

In this chapter we describe, and try to explain, social relationships among female white-faced capuchins. After a brief summary of the general social structure and the theories that primatologists have developed to account for it, we tell the stories of two females, Abby and Tattle, following them over a period of thirteen to fourteen years. At the beginning of our account, Abby is alpha female of her group, and Tattle occupies the lowest position in the hierarchy. Finally, we look at a third female's career to examine the role of alliances in female status competition.

Female capuchins stay in the troop of their birth for their entire life, except when a group becomes too large and splits in two. Even then, however, females retain some of their female companions, and usually the closest female kin remain together. This means that social relationships among females can last for decades, and that middle-aged and old females have the company of their adult daughters and even granddaughters. Male capuchins, in contrast, leave their birth troop during adolescence to seek their fortune, though they often embark on this adventure with a male friend or two. This combination, known as female philopatry (literally, "love of country") and male dispersal, is common in Old World monkeys but unknown in New World monkeys with the exceptions of capuchins and one species of squirrel monkey.

In 1980, primatologist Richard Wrangham first drew attention to the prevalence of what he called "female-bonded" primate species: those in which females are philopatric and form strong bonds, particularly with close relatives, to cooperatively defend access to food and other resources.[2] Because female mammals in the wild can usually transform any additional food into additional babies (that is, shorter intervals between births), they should have evolved to organize themselves in ways that maximize access to food. Male reproduction, in contrast, is limited by access to fertile females, so males should distribute themselves across the landscape in response to the distribution of females. Wrangham argued that female bonding was primarily an evolutionary response to competition between groups over fruiting trees and other high-quality food that occurs in defendable patches. Once two females team up to defend a tree full of juicy figs, for example, the other females in the neighborhood have no choice but to cooperatively defend their access to food as well. Groups will continue to grow until the costs of feeding competition *within* each group become too large to justify further increases in membership. Because genetic relatives have a stake in each other's reproduction, these cooperative bonds are most likely between mothers and daughters, pairs of sisters, and aunts and nieces. Close kin are expected to prefer each other as companions, grooming partners, and coalitional allies.

Wrangham's model inspired a great deal of research, some of which has led other theorists to modify his model.[3] It now appears that the threat of predation, rather than within-species competition for food, forces most primates to live in groups. A larger group has more eyes and ears to watch and listen for approaching danger, more companions to help drive off a predator, and a reduced chance for each individual of being the victim of a successful predation attack (this last is known as the "dilution" or "selfish herd" effect). But although increased group size may enhance survival, it generally reduces birth rates, suggesting that feeding competition within groups is more intense than between-group competition. (If between-group competition were the stronger force, and if large groups usually defeated smaller groups, then females living in larger groups would have access to more food and hence have higher birth rates).

Wrangham's classification of primate species into female-bonded species and female-transfer species now appears too simple, as well, because the female-bonded, or female-philopatric, species vary in key aspects of their social structure. In some species, such as rhesus monkeys, philopatric females form stable dominance hierarchies, usually based on kinship: mothers help their daughters rise to a rank just below theirs. In other species, such as pata monkeys, female-female relationships are more egalitarian. Even among the former, there are differences in how strictly the hierarchy governs access to food and social partners.

High- and low-ranking female capuchins often lead very different lives. Soon after habituating Abby's group in 1990, we realized that Abby herself, as alpha female, enjoyed considerable loyalty and respect. She could usually be found near the center of the group. Normally only the alpha male is entitled to a more desirable feeding spot in fruiting trees. Most striking was the fact that Abby could almost always count on second-ranked Squint for help against any recalcitrant subordinate female or, even more impressive, against any adult male.

Capuchin coalitions are easy to record because they usually involve body contact between the partners, who embrace, stand cheek to cheek, or assume the overlord posture while directing facial threats toward their common opponent. Sometimes coalitions escalate into joint chases or even physical attacks against the target. Of all the female-female coalitions against females that we observed from 1991 to 1993, 38 percent consisted of Abby and Squint, and more than half of their coalitions were directed against third-ranking Nanny. Similarly, about a third of the female-female coalitions against adult males were made up of Abby and Squint, and their most common target was the alpha male, during the reigns of both Curmudgeon and Paul Bunyan. Adult males outweigh adult females by about 40 percent and have much larger canine teeth, so such confrontations entail considerable risk. Abby and Squint spent much of their time in close proximity, but their frequent coalitions were not merely by-products of mutual availability; they often actively sought each other's help. On one occasion, when alpha male Curmudgeon seized a wasp nest that Abby had been moving toward, she ran some fifty meters to Squint, whom she brought back with her to join in an over-lord against Curmudgeon, who nevertheless ignored the two females while consuming his snack. Subordinate males, however, almost always fled from Abby-Squint coalitions. We found a general relationship be-tween female dominance rank and rate of coalitions against males: higher ranking females formed these coalitions more frequently than low-rankers.[4] Female-female pairs that groomed more frequently also joined forces against males more frequently. If they could speak, fe-male capuchins would surely endorse the feminist slogan, "Sisterhood is powerful."

Evolutionary theory and many studies of group-living animals lead to the expectation that closer genetic kin will have closer, friendlier rela-tionships than distant kin or nonkin (Chapter 7). However, in 1991 through 1993, differences in genetic kinship (as revealed to us much later from DNA in fecal samples) had little influence, statistically, on patterns of female grooming or spatial proximity. This was because all the females were so closely related that kinship distance did not offer a useful criterion by which they could discriminate among one another. Abby was the mother of Nanny and Diablita (then an adolescent), and probably the mother (though possibly the sister) of Squint and Wiggy. Tattle, as Abby's grand-daughter or niece, was the "odd woman out"— the female with the most distant average relationship to the others— and this may have partly explained her peripheral position in the group, as described later in this chapter.

The very rare occasions when aggression broke out between Abby and Squint spoke volumes about their relationship. When Abby lunged at or bit Squint, Squint did not retaliate or try to recruit allies as she did when attacked by anyone else. Instead, she gave extremely loud, blood-curdling screams and stared off into space, as if the foundations of her world were being shaken by an earthquake and there was nothing to do but wait until it was over.

Of course alpha rank cannot protect against every kind of reproductive setback that may befall a female capuchin. In May 1993, just four days before we were to leave Lomas Barbudal at the end of our 1993 field season, Abby gave birth to Omni. But instead of providing us with a pleasant send-off, the event ended the field season on a sad note, for the baby lived only three days. Omni apparently had had an umbilical hernia; he remained very weak and could not grasp his mother's fur as capuchin infants normally can almost immediately after birth. During Omni's brief lifetime, whenever Abby moved from one branch to the next, she tried to support him with one arm to prevent him from falling (see Chapter 10). Even in this situation, however, rank did have its privileges: to accommodate Abby's limited mobility, the group stayed within an area about a hundred square meters for the entire three-day period— a striking contrast to the usual one to two kilometers covered in a daily journey. We wondered if the others would have done this for a lower-ranking female.

In December 1996 we returned to Lomas for our first long field season since 1993. I focused my attention on the females and infants of Abby's group and discovered that Abby was still alpha female, and that in fact the surviving adult females from the first study had the same dominance relationships they had had more than three years earlier. Furthermore, when I ranked female-female pairs by grooming frequency, the rankings in this second period, from 1996 to 1997, were very similar to those of 1991 to 1993.[5] It was clear that long-term stability in female social relationships was at least a possibility in capuchin society.

Unfortunately for Abby, her long decline was about to begin. The turning point was the birth of her twins, Albert and Cujo, in March 1997. Strangely, the twins were born ten days apart. At first we found this hard to believe and suspected that Abby had stolen one of the infants from Vandal, who had also been pregnant. However, we found that Vandal was still pregnant. Later we consulted the literature and discovered another case of monkey twins born on different days, and we are now inclined to think that both infants were in fact Abby's offspring. In either case, caring for both of the infants proved too great a burden for her.

Twinning in apes and monkeys (except in marmosets and tamarins, which normally produce twins and have evolved a specialized social support system for caring for them) is rarer than in humans, and for good reason: producing milk for two, and carrying both infants around for months, drains a mother's energy reserves so severely that both the infants' survival chances and the mother's future reproductive potential are sharply reduced. While Abby received plenty of help caring for Albert and Cujo, they died sometime between August 1997 and January 1998, and the whole experience left Abby much thinner than before. Her fur was thinner and scruffier also. She developed a dental abscess that oozed pus for two years and cost her several teeth, so that she had to consume coati pups by sucking on them like Popsicles. In the meantime, an even greater disaster had befallen Abby: Squint, her longtime companion and supporter, had disappeared and presumably died during the same period when Abby lost her twins.

Julie Gros-Louis and two assistants arrived at Lomas in January 1998 to continue collecting data for Julie's research on vocal communication. In February they began to see signs that Abby's position was being challenged. During a routine coalitional interaction in which Abby and two other high-ranking females jointly threatened a juvenile male who had gotten into a tiff with Abby's subadult son Rico, her two comrades began directing an overlord against her behind her back. Then, one day in late April, Abby approached her adult daughters Diablita and Maní and they

Carrying both of her twins is a huge energetic burden for Abby. Photo: Susan Perry.

responded by forming an overlord against her while Diablita hit and pulled at her. Abby tried to distract them by threatening the yearling daughter of low-ranking Jester. Diablita relented and began grooming Abby, but moments later she resumed pulling Abby's tail and bit her. Abby swatted at her tormentors, and the spat finally ended with Diablita grooming Abby. No one joined the squabble on Abby's side. During our 1999 field season, we found that Abby ranked somewhere near the middle of the ten adult females in her group.

By 2001, Abby had become the simian equivalent of a tough and eccentric old lady. Visibly arthritic, she had trouble climbing trees and spent much of her time on the ground. While the rest of the group foraged high in a fruiting tree, Abby would search for fallen fruits amid the leaf litter. During the dry season of that year, Abby's group spent most of its time moving up and down a river that was bordered on both sides by cattle pastures. This ranging pattern was convenient for us during our searches for the group, because capuchins never cross open spaces on the ground—or so we thought. One day Abby bolted across a pasture, oblivious to the grazing cattle, climbed a lonely-looking tree, and spent half an hour foraging in it.

Decrepit though she seemed in some ways, Abby was still healthy enough in other ways. Her final offspring, Bailey, was born in late 1999. Primatologists debate whether any nonhuman primates experience menopause; the emerging consensus is that, among primates, only human females routinely experience a complete shutdown of their reproductive system while their other physiological systems are still in good working order. Of course, some individual female monkeys and apes stop producing offspring several years before they die, but this is a far cry from the long and universal postreproductive period in women. (True menopause is found outside the primates, however, in at least two whale species, the killer whale and the short-finned pilot whale).

Despite her physical frailty, idiosyncratic foraging habits, and low value as a coalition partner, Abby retained a special place in the hearts of most of the group's adult females, especially those who presumably remembered her glory days as alpha. In 2001, the last year we regularly collected detailed data (focal follows) on adults in this group, Nanny and Diablita groomed her only 20 to 30 percent less than they had in 1991 to 1993. Maní groomed Abby almost as frequently as Squint had groomed her ten years earlier. Remarkably, Abby did not groom another adult female for even one second during our 2001 focal observations.

Prevailing theories of the function of primate grooming place heavy emphasis on its value as a commodity and predict that it will be exchanged for return grooming, for coalitional support, or for tolerance from more dominant individuals. Several of our own findings support these views. Besides the link between female-female grooming and coalition formation, we have found that exchanges of grooming between females tend to be balanced within grooming sessions, as if the monkeys are trying to avoid being "cheated."[6] Some of the imbalance that does occur is status-related, for more grooming goes up the hierarchy than down it. Most of the remaining imbalance results from the temporarily enhanced popularity of mothers of young infants: females groom these moms to "buy" opportunities to touch and nuzzle the new baby. Theory also holds that kin are expected to help each other even when there is little or no hope of return favors, but this aid is expected to flow toward relatives with high "reproductive value"—that is, those with bright prospects for future reproduction. In her privileged treatment by her daughters and granddaughters in her dotage, Abby showed us that simple theories have their limits in explaining the behavior of large-brained social mammals. There is no way to know how the other females felt about Abby, but we cannot resist the conclusion that a vestige of respect and loyalty had survived the crash of both her "market value" and her reproductive value.

In April 2002 a senseless crime marked the beginning of the last and most poignant act in Abby's life story. Nanny was killed with a slingshot by one of a group of local men who were illegally cutting trees in the forest. Her eight-month-old son Nobu somehow survived, nursing from other females and quickly improving his ability to catch insects, and the psychological void created by the death of his mother was filled by Abby; the two became close companions. Abby disappeared in February 2003 at a ripe old age, leaving her foster son small for his age but well integrated into the juvenile playgroup. By the time she died, only four of her thirteen-plus offspring were still living, but she left behind fifteen living grandoffspring, ten living great-grandoffspring, and one great-great-grandson.

The life of Tattle, who gave birth for the first time in 1990 (probably at the age of six years, in light of our observations of females of known birthdates) and who during our original study was the lowest ranking female in Abby's group, stands in stark contrast to Abby's. Particularly after a violent male dominance-rank reversal ushered in a period of social upheaval in November 1992, Tattle spent very little time near other adult females. During several hundred hours of focal follows, she formed just one coalition with a female against another female and five coalitions with females against males. Whenever Tattle tried to join the rest of the group's females in a fruiting tree, she was ignominiously chased off. But she did not try this very often, usually remaining content to feed in her own tree some distance from the others. Ironically, perhaps because she faced less feeding competition than the higher-ranking females, Tattle always appeared well fed, if not downright plump. But spending so much time on her own, accompanied only by her young off-spring, created a separate set of problems for Tattle: predation risk. She lost infants in 1990 and 1992, most likely due to predators.

One of our only observations of apparent predation on a capuchin oc-curred in January 1992, when almost the entire group of monkeys was assembled over a vast patch of thorny terrestrial bromeliads (*Bromelia pinguin*, or *piñuela* in local common usage). Suddenly the group began giving the alarm calls that mean "Snake!" Such incidents are fairly com-mon, occurring about twice a week, but this one was unusual in the fre-quency and intensity of the alarm calls and the length of time before the monkeys settled back into their normal activities. Tattle's eight-month-old son Fester was never seen after this incident, and for months after-ward the monkeys gave snake alarms whenever they passed that spot. We tentatively concluded that Fester was taken by a rattlesnake or a boa while foraging for *piñuela* fruit on the ground, but even wearing our snake leggings we were unwilling to chop our way through the dense thorny plants to find out for certain.

Whether coincidentally or as a cause or consequence of her low rank, Tattle sometimes made inexplicable social blunders. On one occasion she was calmly grooming a small juvenile when she suddenly began hit-ting him until he scampered off. Because when this happened we had not yet devised our system for identifying young capuchins, we do not know the full social context of the attack. However, we are almost cer-tain that assaulting another female's child (it was definitely not Tattle's own child) is an impolitic move for the lowest-ranking female of the group.

During my 1996–97 study of females and infants, Tattle was still at or near the bottom of the female hierarchy. Two maturing daughters of Abby—Vandal and Maní—had passed her in rank. She was still socially peripheral among the females, too, but she had intensified a strategy that was already evident in our first observation period: forming close bonds with subordinate adult males. Abby's group contained two of these fel-lows that year, Guapo, whose name (Spanish for "handsome") seemed like a cruel joke now that he had earned the male capuchin's typical bat-tle scars, and Hongo ("Mushroom"), who had immigrated to the group the previous year. During group rest periods, Tattle could usually be found snuggled next to one or both of them. Before giving birth and becoming the recipient of the usual "Can I touch your baby?" attention from other females, Tattle groomed with Guapo far more than she groomed with any female. She groomed with Hongo more than with any female except for Jester, Wiggy's daughter, who shared a spot with Tattle near the bottom of the female hierarchy.

After her son Fink (someone who "tattles" to the authorities is a "fink") arrived on the scene in May 1997, his most enthusiastic admirers were Guapo and Hongo—a striking contrast to the usual pattern in which adult and adolescent females show far more interest in another's infant than males do. On one occasion Fink was on Hongo's back, gargling to him, when a chorus of "other monkey" alarms rang out: a hostile capuchin troop was nearby. Hongo ran off to join the battle, still carrying Fink, while Tattle ran after them in a panic. Female capuchins may well prefer males that are both brave warriors and devoted babysitters—but not at the same time.

By 2001 Tattle was part of a growing clique of low-ranking females that included Jester and Eldritch, Tattle's nine-year-old daughter. These three weren't nearly as cohesive as Diablita, Vandal, and Maní, the "in-crowd" at the top of the female hierarchy, grooming each other only about one-sixth as frequently. This is a common finding in female-bonded primates, and it makes sense if the function of social bonding is to cultivate valuable allies. If you have a low rank, your fellow low-rankers are not as valuable to you as a high-ranker would be—but high-rankers are unlikely to see much value in you, so you may not get much opportunity to schmooze with them. Groucho Marx expressed the nub of the problem: "I don't care to belong to any club that will accept me as a member." This intuition forms part of the basis of primatologist Robert Seyfarth's influential model of grooming in female primates.[7]

As explained earlier, according to this model more grooming should go up the hierarchy than down. Since time for grooming is limited, females will compete for grooming partners. High-rankers will win the competition to groom with fellow high-rankers, middle-rankers will have to content themselves with middle-ranking grooming partners, and the low-rankers will be stuck with each other. This means that most grooming will take place between females of similar rank. These females will often be kin, because in many female-bonded species (though not white-faced capuchins; see below) mothers consistently support their daughters' ambitions to rise in rank until they reach a point just below them, creating a situation in which close kin hold similar dominance ranks. The females of high-ranking families will be drawn together by both kinship and the desire for valuable allies, whereas low-ranking females will have only family ties to bind them; thus high-ranking families will be more closely knit than low-ranking families. In a recent analysis, using sophisticated statistical methods to combine data from different studies, Gabriele Schino showed that Seyfarth's model is generally consistent with what we actually see in group-living female primates.[8] Among the females of Abby's group, we have found that subordinates groom dominants more than vice versa (though this effect is fairly weak) and that high-rankers groom each other more than low-rankers do.

The third prediction of the Seyfarth model, that closely ranked females will groom each other more than distantly ranked females, was supported in the early years of our study, while Abby's group was small and most of the females were close kin. But as the group grew and average relatedness fell, kinship acquired increasing influence over grooming patterns, while rank lost influence. Females now had a greater variety of kin types (from mothers and daughters to distant cousins) from which to choose when selecting grooming partners. By 2001, when the group contained ten adult females, closer kin groomed each other much more than distant kin, while rank distance had no effect on grooming frequency.

We also found that kinship affected female-female coalitions against females. We examined all observed female coalitional triads (female A successfully solicits female B against female C) and found that, out of nineteen triads in which B was not equally related to A and C, seventeen were coalitions of closer kin against a more distantly related female. In addition, females did not always support their juvenile and adolescent daughters' efforts to rise in the social hierarchy. Taken together, our results show that the Seyfarth model has only limited applicability to white-faced capuchins.

In 2002–2003, immigrating males brought chaos and bloodshed to Abby's group (see Chapter 8). Two of Tattle's grandoffspring, consecutive infants born to Eldritch, were victims of infanticide. Fortunately for Tattle, her most recent offspring, Toulouse, was already weaned, and was therefore no longer a tempting target for infanticidal males, but an entire cohort of the group's infants was wiped out in 2003. In April 2003, Tattle's two subadult sons, Solo and Fink, emigrated at the same time as six other young males. By mid-2003, Tattle's clique of low-ranking females, their young offspring, and a changing cast of adult males had begun spending more and more time far from the rest of Abby's group. Late in the year we christened this newly independent group "the Flakes" and were somewhat surprised, given her lack of social skills, to discover that Tattle was its alpha female. Susan and Hannah, who had known Tattle well as a subordinate, were amused to see her in the middle of the new group, asserting herself and cozying up to the new alpha male. She became pregnant again, and we cheered her on in her new role as grand matriarch of the Flakes. But her glory days lasted just a few months, and ended with her disappearance in February 2004. We never found Tattle's remains or discovered what had happened to her, but we suspect that she was killed by a farm dog at a place where she often descended to the ground to eat mangoes.

Coalitions are ubiquitous in the lives of female capuchins, and they usually reinforce the existing hierarchy. However, these facts by themselves do not tell us how female capuchins acquire high rank. Are dominants born with a silver spoon in their mouth, coasting easily to high rank with the help of dominant kin and the deference of the low-born? Or do they fight their way to the top, attracting allies who are drawn by their prowess? We have seen that kinship affects alliance patterns, but how do females discriminate, if they do, among equally closely related females in allocating their political support?

The career of Diablita, the most socially ambitious female in Abby's group, offers some answers to these questions. As an adolescent in 1992–93, she had a tense, ambivalent relationship with the much older Wiggy, as the following data excerpts show. Up until that time Wiggy had been the fourth-ranking female.

June 15, 1992. Diablita approaches Wiggy and grooms her for just one second before presenting for grooming. Wiggy responds by presenting for grooming herself. Just then Abby approaches to five body lengths of them. Diablita directs a threat face to Wiggy, pushes her, pulls her tail, and threatens her again. Abby approaches to one length, Diablita threatens Wiggy again, and Wiggy leaves to a distance outside five lengths from Diablita and Abby.

August 24, 1992. Diablita and Wiggy are play-wrestling, something quite unusual for adult females. Suddenly the play turns into serious ag-gression: Diablita bites Wiggy's hand and chases her, screaming. Abby threatens the fleeing Wiggy, and Squint and juvenile male Hobbes get into an overlord posture directed against her.

Throughout 1992 and early 1993, Diablita and Wiggy were frequent grooming partners, and Diablita was very solicitous toward Wiggy's year-ling son Yoyo. Yet Diablita also launched frequent unprovoked attacks against Wiggy, almost always receiving the active or tacit support of Abby, Squint, or Nanny—or some combination of them—who out-ranked Wiggy. By the time we left the site in May 1993, Diablita, then pregnant with her first offspring, was unquestionably ranked above Wiggy.

To this point, Diablita's story resembles literally hundreds of cases re-corded during the past forty years by observers of baboons, macaques, and some other female-philopatric societies. Rhesus and Japanese ma-caques, in particular, were said to consistently follow the "younger-sister ascendancy rule," according to which a female's mother and other kin would help her acquire a rank just below her mother and just above her next youngest sister.[9,10] The pattern was explained as resulting from mothers' supporting their smaller, and hence more vulnerable, juvenile and adolescent daughters during squabbles with an older sister until the older of each pair of sisters accepted her new lower status.

However, the younger-sister ascendancy rule turns out to be largely confined to macaque populations being provided with extra food by hu-mans. Primatologist David Hill has argued that concentrated artificial food sources, such as the hog feeders full of monkey chow used at the Cayo Santiago rhesus macaque colony in Puerto Rico, promote more se-vere aggression than occurs over more widely dispersed natural foods.[11] The resulting high risk of injury to juveniles, in turn, leads mothers to support their younger daughters more vigorously and consistently than in wild populations. After we learned the genetic relationships among our females, we were able to look at the effects of kinship structure on dominance ranks in white-faced capuchins. Diablita's career illustrates the lack of a clear-cut relationship between them.

Four years after rising above Wiggy, Diablita, still ranked fourth, tried again to climb the hierarchy with help from those at the top, but this time the outcome was different. Her confrontations with third-ranking Nanny, who was also older than Diablita, peaked in early February 1997:

**February 1, 1997.** Squint and Diablita run up to Nanny. First Diablita and then Squint lunge at Nanny. Then they both chase her, Diablita in the lead. Nanny chews her tail, a sign of extreme distress that we observe only in victims of coalitionary attack. Diablita continues to chase Nanny and breaks a branch near her. Nanny chews her tail again, Diablita glares at her, and Nanny responds by glaring at the human observer, a common form of attempted distraction by capuchins that are on the losing end of fights. Squint again threatens Nanny, and Diablita resumes chasing her.

**February 3, 1997.** Diablita approaches to ten lengths of Nanny and threatens her. Nanny threatens an observer, but then threatens Diablita, and they exchange threats for two minutes. Squint approaches Diablita, grabs her hand, and sniffs it. Diablita threatens Nanny twice more, and Nanny responds by chasing the previously uninvolved Eldritch, who flees. Diablita and Squint follow, both threatening Nanny. For a moment it looks as though Nanny's ploy has succeeded, as Diablita, Squint, and Abby (who has just arrived) all threaten the hapless Eldritch. Then Squint gets into overlord posture over Diablita against Nanny, and Abby also threatens Nanny. Squint gets off Diablita's back and begins an overlord over Abby against Nanny. But then Squint changes her mind and begins threatening Diablita, who of course is still threatening Nanny. Now Squint alternates her gaze between Nanny and Diablita, seemingly unable to decide whose side she is on. Diablita cowers to Squint, who approaches and mounts her. After dismounting and alternately threatening Nanny and Diablita several more times, Squint stands by as Nanny flees the scene chased by Diablita.

Two days after this dramatic display of Squint's ambivalence, the frequency of Diablita-Nanny conflicts fell sharply, and from that point we observed no other monkey supporting Diablita against Nanny. On

March 19, Squint unequivocally supported Nanny against Diablita, and from then until the end of the field season in June, Nanny was also the clear winner in all one-on-one contests with Diablita. The existing hierarchy had held firm; Diablita's rise would have to wait until the following year.

When Diablita finally took the alpha position a few years later, she again violated the younger-sister ascendancy rule, because two of her younger sisters, Vandal and Maní, remained below her in rank. Vandal took the alpha position in 2004 (in a surprising alliance with the new alpha male Tranquilo), by which time both she and Diablita were approaching middle age, while Maní, who is younger than Vandal, remained below Diablita in rank. Reviewing the careers of all the Abby's-group females up until 2001, we found an average difference of one to two rank positions between each female's actual rank and the rank she would be expected to hold if the macaque younger- sister ascendancy rule applied to white-faced capuchins.

After our 1997 field season, we were convinced that we had an interesting story to tell about long-term trends in capuchin female-female relationships, but we wondered whether the Lomas Barbudal capuchins were typical in this respect. As described earlier in this chapter, resource distribution and feeding competition are thought to play major roles in shaping the social lives of female primates. Our study area is crisscrossed by permanent streams, giving our monkeys easy access not only to drinking water but also to dry-season food sources such as the fruit of the *Sloanea terniflora* tree, which grows only in close proximity to a permanent water source. Not all capuchins in northwest Costa Rica are so lucky. Researchers working at the poorly watered Santa Rosa National Park, fifty kilometers from Lomas, told us that by the late dry season, their monkeys were reduced to subsisting largely on ant larvae harvested from acacia trees (see Chapter 3), a fallback food rarely consumed by our capuchins.

Perhaps the harsher conditions at Santa Rosa had selected for different patterns of female-female relationships. To find out, we combined our data sets with those of Lisa Rose, who studied the Santa Rosa capuchins from 1991 to 1997. To extend the timeline of the Lomas females' careers, we also collaborated with Julie Gros-Louis, who had conducted her own research at Lomas during parts of 1996 to 1998 while we were away from the site, teaching at UCLA.[5]

All three troops observed by Rose experienced social upheavals among adult females. One alpha female lost her rank to a coalition after her most frequent grooming partner disappeared, just as Abby fell from power after Squint's presumed death. In two other cases, the disappearance of alpha females touched off power struggles among former subordinates, but the extent to which these contests were decided by coalitional support was unclear. The Santa Rosa data showed neither the frequent hierarchy-reinforcing female-female coalitions observed at Lomas, nor any hint of preferential up-hierarchy grooming, nor the strong association between grooming and mutual coalitional support in comparisons among pairs of females.

The reason for these differences between sites may lie in the much higher death rate among adult females observed at Santa Rosa compared with Lomas Barbudal during 1990 to 1998. It is probable that monkeys adjust their social tactics to prevailing circumstances. Frequent deaths in small groups, and consequent social upheaval, may reduce the payoff from investing in relationships with dominants (they may not be around for long) and may place a higher premium on forming shifting, opportunistic alliances. We do not know the Santa Rosa females' kinship relationships, but it is possible that high death rates reduce the probability that any particular female will have close female kin who can be relied on for support.

At both Santa Rosa and Lomas, especially in smaller groups, dominance struggles among females are often decided by one-on-one contests rather than political maneuvering. Declining physical condition in middle-aged females sometimes precipitates falls from high status; we do not see elderly matriarchs retaining alpha rank by riding their reputation. From all this we conclude that coalitions and kin networks are often a central element in female capuchin politics, but that individual fighting ability and plain luck play important roles too.

# CHAPTER 7

# Curmudgeon:
## The Career of an Alpha Male

**January 19, 1992**     I am doing a focal follow of Curmudgeon, the alpha male of Abby's group. He is swaggering down a branch, with his hair fluffed out as usual and a scowl on his face. His face is so scarred and twisted from fights and mottled by age spots that it is no longer possible for his facial expressions to look benign by human standards, even when he is in a relatively good mood. An infant rushes toward him, with his hair standing on end and penis erect, repeatedly making the horrible, loud gargling sound infants direct toward an alpha male. The youngster occasionally pauses to squirm excitedly and scratch himself before resuming the gargling noise. He gazes fixedly at Curmudgeon, kneads his tail tip thoughtfully, and then cocks his head and protrudes his pink tongue, curling it down to his chin. Finally, still gargling, he musters the courage to approach his idol. He cautiously touches Curmudgeon, who has just flopped down on the branch for a rest. The infant grooms Curmudgeon perfunctorily and then attempts to suck from his nipple.

Abby saunters over, sits down in contact with Curmudgeon, and begins to groom him. Seconds later, Nanny approaches, gently nudges the infant out of the way, and also starts to groom Curmudgeon. Her infant also runs over, touches Curmudgeon, and begins to groom him while making gargling noises. Nanny and her infant get up and leave after a minute or two. Abby stops grooming, and Curmudgeon raises his head, glowers at me, and threatens me by baring his teeth. Abby leaps to attention, following Curmudgeon's gaze, and also threatens me by uttering a little squeak along with her threat, to show her support for the alpha. Abby's own infant rushes over, gargles, and touches Curmudgeon, and both Abby and the infant fervently gargle and groom him for several more minutes.

This is a typical scene in the life of a capuchin alpha male. Wherever he goes, he is followed by an entourage of adoring fans who groom him, gargle at him, and eagerly form coalitions with him. He is so popular that he can hardly find a moment's peace. In contrast, subordinate males spend most of their time on the periphery of the group, foraging alone.[1] Subordinate adult males do have friends, most often juvenile males, who are eager to engage in rough-and-tumble play with them. But even though they usually dominate all females in the group and therefore occupy ranks just below that of the alpha male, subordinate males do not get nearly as much respect as the alpha male does. In fact, they are the group's most popular scapegoats.

At the slightest provocation, or sometimes without any provocation at all, adult females and juveniles will suddenly scream at a subordinate male and solicit the alpha male's aid in evicting him from the tree in which they are feeding. Once a large enough coalition has formed, the subordinate male is forced to flee from the tree, and from a distance will nervously chew his own tail while watching the feast from the sidelines. It must be galling to sit there, hungry, waiting for the juveniles and low-ranking females to finish gorging themselves on the best fruit. If the subordinate male tries to sneak back into the tree, the juveniles and females sound the alarm with a squeaky threat or scream, and the male is driven out again to wait his turn. In a one-on-one contest, of course, the adult male could easily win a fight against any of the persecutors, but in capuchin society allies count for a lot. Subordinate males rarely have the political support they need to accomplish their goals.

**January 27, 1992**     I am following Abby's group as the monkeys lazily meander down the banks of a river during the noon heat, foraging for insects. Suddenly the authoritative voice of Curmudgeon pierces through the soothing sound of water bubbling over rocks, as he announces the discovery of a boa constrictor hiding in the grass. He stands on the ground, hair erect, just a couple of feet from the snake, repeatedly barking out alarm calls, bouncing up and down at the snake and glowering at it. He is soon joined by a number of group members, who gather round to stare at the snake and alarm call just as Curmudgeon is doing. Several pairs of monkeys stack on top of each other in the overlord position, baring their teeth at the snake. After a few minutes of outraged demonstrations against the boa constrictor, Curmudgeon retires to a nearby tree branch for a rest, leaving the others to persecute the snake. The rest of the group tires of this activity after a while, and everyone settles down to groom or rest.

The two subordinate adult males, Paul Bunyan and Guapo, sit down next to each other, and Paul begins to groom Guapo. Male-male grooming is a highly unusual event: I had never seen any male groom with Curmudgeon, and this pair of subordinate males had been recorded as grooming at the low rate of only 1.6 seconds per hour. Curmudgeon spots Paul and Guapo engaged in this rare moment of intimacy, and he immediately leaps to his feet and rushes over to within a meter of the snake, although it has not moved at all since being discovered and all the monkeys have lost interest in it. Curmudgeon alarm calls furiously at the snake as if it has just committed some new atrocity, and shakes branches at it, glancing at Paul and Guapo every couple of seconds. It seems clear that Curmudgeon is trying to divert their attention from grooming each other by reminding them of the snake. The ploy works, and both males immediately stop grooming and rush to Curmudgeon's side. All three of them harass the snake for a while, and then Curmudgeon and Guapo wander off together, leaving Paul alone with the snake.

In general, male-male relationships in capuchins are tense and somewhat ambivalent, in contrast with the much more relaxed and stable female-female relationships.[2] Males are not avid groomers in general (though they all love to receive grooming, they balk at reciprocating). With rare exceptions, they seem slightly aroused and nervous in one another's company, and they are always monitoring one another's activities. Even play and grooming, when performed by pairs of adult males, tend to have a different quality to them. Their motions are stiffer, more exaggerated and more frenetic, giving the interactions a slightly hysterical quality, in contrast with the more relaxed nature of grooming between females. During the first few years of my research, all male-male relationships had this tense quality to them, though I did eventually come to know some pairs of males who seemed more comfortable with each other. Alpha males typically become nervous when they see subordinate males affiliating with other subordinate males and rush over to break up the interactions.

Despite their uneasy relationships with one another, all capuchin males can agree on some matters: for example, that snakes are bad news; ditto humans, dogs, cattle, and practically every other type of animal that they encounter in the forest. It is common for male capuchins to head off potentially unpleasant interactions with each other by collaborating in harassing some other creature. Primatologists are particularly useful for this purpose, and we like to think that we have alleviated the stress level of many a capuchin monkey simply by being conveniently present for scolding when two male capuchins wished to forage in the same tree. Typically, when a subordinate is foraging and a dominant male blusters into the tree as if ready to supplant him, the subordinate (who has been foraging peacefully just a meter from my head and totally ignoring me) will suddenly spin about and threaten me, headflagging rapidly over his shoulder to recruit the approaching male. Both males will then direct threat faces and vocal threats toward me for a while, as if reaffirming their common bond against dangerous creatures, and then settle down to feed together peacefully.

Most of the capuchin alpha males I have known have been control freaks, constantly monitoring, commenting on, and manipulating the relationships of the subordinate males in their group. In most groups, for example, the alpha male responds to wheeze dances between subordinate males by running toward them and emitting intense vocal threats. But there has been one highly notable exception: Pablo, the alpha male of Rambo's group, has quite a relaxed personality and typically ignores such events. Pablo permits frequent grooming and is an avid groomer himself (for a male), not only of females but also of juveniles (most of whom are his offspring) and his male allies. In addition, Pablo, unlike other alpha males, permits male-male sexual interactions among the subordinate males in his group.

It is amazing what a difference the personality of the alpha male can make in the relationship dynamics of an entire group; when switching from Abby's group to Rambo's group, I always feel like I am entering a different culture. The males in Rambo's group generally seem comfortable associating with one another in a relaxed way and are highly cooperative. As a result, the Rambo males are able to form a united front against enemy males that is frighteningly effective.

Pablo's technique for ruling is so different from that of the other alpha males that we have been expecting him to be overthrown for many years now; after all, most capuchin "nice guys" seem to meet a sticky end (see Chapter 9). But Pablo has been solidly in place since 1996, the year we started studying the group, and probably since 1991 (judging from his monopolization of paternity in previous years), and we have seen no attempts to unseat him. Perhaps because the male bonds have always been so strong in Rambo's group, males sometimes sexually coerce females in this group, while male coercion of females is virtually unheard of in the other capuchin groups I have studied. Female-female bonds in Rambo's group are relatively weak compared with those of Abby's group during Abby's tenure as alpha female.

Another way in which the more typical capuchin alpha males maintain control over subordinate males is by being fickle in their coalitionary support. In addition to breaking up coalitions and affiliative interactions between subordinates, they are inconsistent regarding which of the two males they support, sometimes switching multiple times in a day, such that the subordinate males have no idea where they stand with the alpha male. Probably owing largely to this particular alpha-male tactic, we cannot discern linear dominance relationships among subordinate males in many cases. These males do not have much of a chance to sort out their relative status because the alpha male prevents or interferes with their attempts to form stable relationships with other males. Despite the alpha male's fickleness, the subordinate males are unwaveringly loyal to him, always siding with him against other males, whether they are from their own group or from other groups. The alpha male is everyone's preferred coalition partner. Females, too, reliably choose the alpha male as an ally against subordinate males. When there is a switch in the identity of the alpha male, the females instantly change their allegiance to the new alpha.

For the first two and a half years of my study, Curmudgeon was the alpha male of Abby's group and group dynamics were completely stable. It was

Curmudgeon (on top) and adult female Wiggy jointly threaten observers.
Photo: Susan Perry.

hard to imagine anyone else being in charge. Though his popularity with the adult females was obviously waning a bit by the end, the females were not giving any other males any encouragement, and they continued to use the subordinate males as scapegoats, soliciting Curmudgeon's aid against them. Mudge (as he was called for short) vocalized almost constantly, and his voice had such a ringing, authoritative quality that I could always tell where he was, even when I was out of view of him. He was unquestionably the most self-confident monkey I had ever known, and his hair stood on end at all times, making him look even larger than he was. He was such a charismatic and amusing individual that we gave him the nickname "His Gloriosity," and he became the subject of numerous silly monkey songs, such as "Glorious Mudge" (a parody of the Flanders and Swann song "The Hippopotamus," with its chorus, "Mud, mud, glorious mud"). It was as if he had his own personality cult, including both humans and capuchins.

Imagine our surprise, then, on November 10, 1992, when, after returning from a long trip to the Costa Rican capital, San José, where we had gone to renew our research permit, we found Curmudgeon wounded, alone, and giving "lost" calls.[3] It was virtually the first time he had been seen alone, and only the second time he had been wounded, since we had known him. He had puncture wounds on his right leg and left arm, as well as gashes on the sole of his left foot and on his shoulder. With both legs nonfunctional, he was dragging himself up trees by his arms. He looked deflated, with his hair lying flat for the first time ever, and he seemed miserable and terrified. He didn't even offer his usual loud commentary while eating. When he lost called, his voice cracked and trembled, each syllable ending with a whiny whimper, in stark contrast to the confident, steady quality his voice had had before. It was truly heart wrenching to see this magnificent animal laid so low. It was also agony to think that we had missed what must have been the most socially significant event of the past two years—during a trip to renew our permit!

The next day I arrived at Mudge's sleeping site at 5:00 A.M. He awoke ten minutes later and began producing the same pathetic, whiny lost calls he had made on the previous day, whimpering between calls. Finally he heard an answering lost call. He paused, as if contemplating his next move, and then responded with more lost calls. You would think that a monkey in his condition would want to avoid contact with any other monkeys until he was fully healed, because he was clearly in no condition to fight again. But perhaps the loneliness was just too much for this highly gregarious monkey; he was probably longing to find out what everyone else was doing and to have some company. Curmudgeon began producing a call I had never heard before. It was more like an alarm call than anything else, but it had a whiny, more tonal quality to it than a typical alarm call and was quite high-pitched, with a sort of hysterical-sounding edge to it. Mudge moved toward the answering monkey and continued to produce these strange calls. I couldn't stop marveling over the odd quality of his voice; it really sounded like he was having what in humans would be termed a nervous breakdown.

Most of the calls he produced were recognizable, but all had that out-of-control, jittery quality to them. Curmudgeon stopped moving and started doing burst-trill-bray sequences, the calls he used to use to initiate group movement. It seemed that he wanted the group to come to him. His voice began sounding more controlled the closer the answering monkeys came.

Finally, after half an hour of vocal negotiation, Paul Bunyan arrived on the scene, and the two monkeys burst into a wheeze dance. Curmudgeon couldn't perform all the elements of the dance with his injured foot, but he flailed and twisted. The wheezes were especially loud and frenzied sounding. Both monkeys looked terrified of each other and kept their distance. Paul had also sustained some rather severe wounds in the face, but he was in otherwise fine condition, and his hair was standing on end so that he looked much larger than I had ever seen him look. Paul attempted to enlist Curmudgeon to join him in threatening us by alarm calling and headflagging to him, but Curmudgeon refused to cooperate. They simultaneously burst into a wheeze dance once more, this time grunting as well as wheezing. Both of them flailed, reared back as if to do back flips, spun about, and hopped while dancing. They looked like nervous wrecks, anxious about approaching more closely, but not exactly wanting to retreat, either. When they finished dancing, both resumed alarm calling. Alarm calls are typically directed toward potential predators and occasionally toward capuchins from other social groups; this was the first time I had seen a monkey direct alarm calls to a member of his own group. After about nineteen minutes, Paul left to return to Abby's group and I followed him.

When we reached the group, it was immediately apparent that Paul Bunyan was the new alpha male. Formerly he had skulked phlegmatically on the edge of the group, rarely interacting with anyone. Now he strutted about in the center of the group, hair standing on end at all times, showing off constantly. He broke branches, threatened me at every opportunity, and solicited the females' and Guapo's aid in menacing me. Three of the females had wounds, most likely stemming from the fight in which Curmudgeon was ousted. It is extraordinarily rare for females to have wounds. I tried to imagine the fight. A single monkey could not possibly have inflicted so many wounds on Curmudgeon. It must have been the case that some held him down while others bit him.

Over the course of the next three weeks, Curmudgeon made several failed attempts to rejoin the group. Each time he was driven away by Paul and the females. The juveniles and infants (whom we now know to be Curmudgeon's offspring) were far more amiable, but the adult females (aside from Wiggy) had made a complete shift in allegiance to Paul. Paul was now the only male they groomed and the only male they gargled to. Guapo had emigrated from Abby's group after a few days of nervously avoiding Paul, so Paul was now the only adult male in the group.

The tables turned somewhat on November 28, nearly a month after Curmudgeon's downfall. At daybreak, Curmudgeon, Guapo, and a third male previously unknown to us (whom we christened Ichabod because of his scrawny build, big ears, and nervous demeanor) all attempted to immigrate. Paul lost his cool when confronted with this new problem. Although the females had been supportive of his attempts to drive out Curmudgeon, they had fought with him several times in the past weeks and even wounded him. Normally the subordinate resident males would help the alpha male fight off an intruder, but since Paul had been mistreating his former allies for a month, we weren't terribly certain he would get the backing he needed to force out the new young challenger, Ichabod. Perhaps all three males had an easier time entering the group because their simultaneous arrival rattled Paul so much. As the sun came up and Paul looked about the group, assessing the situation, he fear-grinned spasmodically, wheeze danced with Curmudgeon, and emitted several alarm calls. He headflagged to Curmudgeon to request assistance in threatening Ichabod. Curmudgeon, who seemed more poised than I had seen him since he had been deposed, coolly ignored Paul's request for several minutes before finally joining Paul in the threat. Paul stayed as far away as possible from Ichabod, but Curmudgeon and Guapo were curious about him and even engaged in wheeze dances with him. Curmudgeon and Guapo also had an enthusiastic reunion with each other, consisting of wheeze dances and sex. Curmudgeon had some tiffs with Paul later in the day in which they threatened each other with their hair on end.

Over the next few days Paul concentrated his energies on disrupting Curmudgeon and Guapo's attempts to form an alliance. Paul ignored Ichabod as much as possible, though he did vocally protest his dances with the other males, from a great distance. For a while it looked like

Curmudgeon might be able to make a comeback as alpha, but Paul gradually gained confidence and managed to keep his seat. Curmudgeon slowly became accustomed to his new role as a subordinate male, though it seemed to be a difficult transition for him. On some mornings he still woke up and produced the same whiny alarm calls we had heard during that first week after his fall from alpha rank. He also did this sometimes while watching Paul being groomed in the center of the group by the females who had once groomed him. Whenever Curmudgeon had these noisy outbursts, Paul and the females would threaten him. Most of the time, however, Curmudgeon managed to stay out of trouble. For someone who had been so flamboyant at all times, he became remarkably good at sneaking about at the edge of the group and diving silently into a bush whenever any commotion began for which someone might be seeking a scapegoat.

Clearly he had not forgotten how to be an alpha male, however. On those very rare occasions when Paul left the group to accompany a sexually receptive female, Curmudgeon would take immediate advantage of the situation. These episodes were heartwarming and hilarious, because it was as if Curmudgeon was trying to pack a months' worth of alpha-male displays into an hour's time. For example, on March 13, 1993, Nanny, who was in estrus, or "heat," made a mad dash away from the group, and Paul, Ichabod, and Quizzler followed her. Ever vigilant, Curmudgeon and Guapo watched them go. A couple of minutes later, Curmudgeon swaggered up to me, hair standing on end for the first time in months, and slapped me on the toe. Normally he ignored me completely, so I was surprised by this outburst. He produced a deafening sequence of intense gargles that I was sure would bring Paul back to the group to investigate, but he and the "estrus entourage" must have been quite some distance away by then.

Curmudgeon leaped into a small tree in which two other monkeys were foraging and completely destroyed it in an astounding display of branch-breaking bravado, so that the other occupants of the tree fell to the ground and turned to threaten him indignantly. Curmudgeon threatened me and urine washed, splashing urine everywhere. Then he hopped up and down repeatedly, as if on a pogo stick, reaching astounding heights. He then wrapped his tail around the trunk of the tree he had just felled and attempted to drag it behind him as he charged away.

It proved too heavy to drag far, but he did succeed in making an incredible amount of noise. Curmudgeon broke off a smaller portion of the tree, produced some intense gargles, and dragged that off through the leaf litter with his tail. He and some of his offspring threatened me for a bit, and then he dragged more branches with his tail. As two of his offspring approached him, trilling, he broke a branch over my head and splashed some more urine about. He began charging madly through the group, vocalizing at the top of his lungs and breaking everything in sight. Twice he attempted to drag away branches that were attached to something, and he ended up hopping up and down in a very silly manner as the branches pulled his tail back.

Taking to the trees again, he whacked me on the head with a branch and urine washed once more. Back on the ground, Mudge ran up to a huge log and pushed it with all his might, but it barely budged. Then he took a flying leap toward me and landed in front of me, looking like a giant fluffball. He seized a huge branch with his tail that must have been five times his weight, and dragged it through the leaf litter in a truly spectacular display; this time he had gauged the properties of the object correctly. He produced more intense gargles (muffled by some food he had just stuffed in his mouth) and heaved an enormous chunk of wood covered with hundreds of daddy longlegs into the river, making a satisfying splash. Mudge urine washed and shook branches a few more times before turning to the tree occupied by Abby. She threatened him as he came flying through the air toward her, but he nonetheless proceeded to break the entire tree in another impressive display.

Next Curmudgeon turned his attention to a herd of unsuspecting feral cattle that were lazily lumbering toward the group of monkeys. He zoomed right up to a cow and whacked it with a branch. Then he descended to the ground and chased the cattle, forcing a stampede. When they stopped running, he climbed into a tree above them once more and urine washed, splashing urine down on the cattle multiple times. He began alarm calling and pouncing on the cattle from above, whacking them with branches and urine washing some more. This cattle-herding project kept him occupied for quite some time. Curmudgeon continued to harass cattle, urine wash, break branches, and dash madly through the group, making as much noise as possible, until he heard a wheeze dance in the distance at about 12:30 P.M. He ran toward it, doing intense gargles, and then stopped dead in his tracks as he realized it was Paul Bunyan returning. When he caught sight of Paul taking a bath in the stream, he stopped vocalizing and his hair started to fall flat again. It was like watching a balloon deflate; suddenly Curmudgeon looked much smaller, as if all the stuffing had been knocked out of him. The fun was over, and it was back to the life of a subordinate.

The coup against Curmudgeon and the subsequent changes in male relationship dynamics in Abby's group were to me the most fascinating aspects of the data set from my original field study, in 1991 to 1993, and during our time back in the United States I was eager to return to Lomas so I could find out more about how male capuchins negotiate aspects of their social relationships. Were there subtle details I was missing, such as long-distance vocal communication or visual monitoring? Did the behavior of subordinates change substantially when they were being watched by alpha males?

Obviously I was going to need to study more than one group of monkeys to find this out. I had already selected Rambo's group as my next study group, since it had a large number of males and a home range overlapping with that of Abby's group. Julie Gros-Louis worked hard on habituating Rambo's group for four months in 1996, and Joe and I returned to Lomas in December 1996, four months after Julie left, to continue the habituation process, bringing several field assistants with us. We would continue to study Abby's group as well, of course.

Julie had warned us that Curmudgeon, Paul Bunyan, Ichabod, and two other males named Hobbes and Kola had disappeared from Abby's group earlier in the year, so we were prepared for that shock. In the months between Julie's summer field season and our return to the field, I was so anxious to find out what these missing males were doing and where they had gone that I had dreams in which I was desperately roaming the forest looking for them. In the most memorable of those dreams, I found Ichabod wandering the forest alone and asked him what he had been up to, though not, of course, expecting a response. To my delight and astonishment, this dream Ichabod responded by bursting into an operatic aria in which he boasted about his seduction of several capuchin females. (This is what happens to primatologists who listen to opera while entering data for days on end—their Don Giovannis become arboreal and furry.)

Quizzler, a subadult natal male, was alpha of Abby's group when Julie was there in the summer of 1996, but he had vanished by the time we arrived, as had the adult female Wiggy and two infants. The new alpha male was none other than Ichabod! We did not at first recognize him, because he had filled out, obtained new scars, and become vastly more self-confident and aggressive in his new role as alpha male, and we named this "unfamiliar" male Ramades. But genetic testing and comparison of the patterning of freckles on his face in a sequence of photos later confirmed that Ramades and Ichabod were one and the same. It is shocking how much weight a male can gain quickly when he is suddenly free from the stress of being a subordinate.

Ichabod/Ramades was a rather disappointing alpha male: he was impulsively violent and downright rude to his groupmates during feeding time, and fairly antisocial. He had been unusually unpopular with the females at the time of his initial immigration in 1992. For months after he first joined the group, Ichabod was brutally attacked at least once a day in nasty fights that would end with a ball of monkeys (several females plus himself) falling out of a tree in a writhing, screaming, biting mass and scampering away into the undergrowth. Undaunted, Ichabod doggedly returned each time, eventually wearing the females down. As alpha male from late 1996 to 1998, he was groomed by the adult females and sired two offspring, but he never did achieve the relaxed quality in his social interactions with females that past alphas had had. Our field assistants thought that he was a perfect example of a capuchin sociopath, and we all rooted for Guapo to become alpha male in his place. But both Guapo and the only other male, Hongo (a young male immigrant about Guapo's age), seemed thoroughly terrified by Ichabod and lacking in female support, so this prospect seemed unlikely.

Disgruntled by the sad state of affairs in Abby's group, I desperately wished to see the charismatic Curmudgeon and kept my eyes peeled for him whenever we encountered new monkey groups or all-male groups in the forest. After four months passed without encountering any of the missing males, we began to lose hope of locating them. We described them to our field assistants in excruciating detail, and showed them pictures, in case they should encounter them in our absence. In April 1997, while the assistants were away for a few days of well-earned vacation, we searched around the nearby rivers and streams for Curmudgeon. Although we encountered many new groups, none of them contained monkeys we knew. Sadly, we resigned ourselves to the idea that Curmudgeon was dead and tried not to think about him anymore.

Just days after our failed "Curmudgeon quest," I was doing a focal follow of Q, a young male in Rambo's group who had been named by a Star Trek fan, while field assistant Todd Bishop followed Rambo, the feisty young adult male for whom we had named the group. Three hours into our follows, Todd lost track of Rambo and began frantically searching the perimeters of the group for him. Todd was one of the most zealous field assistants I had ever had, and he would risk his life rather than lose a focal animal. So I was surprised when he still hadn't found his monkey thirty minutes later. I was even more puzzled when he reappeared with an excited gleam in his eye and quietly requested that I come help him identify a monkey. I looked at him like he was crazy and reminded him that I was in the middle of a follow. He just smiled and said, "I really think you should take a look at this monkey. He's a really old guy, and he's pretty tame. He didn't alarm call or threaten me when he saw me, and he barely moved out of my way when I went near him." Since Todd had never shown bad judgment before, I abandoned my focal animal and followed him. Who could this be?

Todd led me downstream and whispered, "Behind that tree!" I sneaked cautiously around a huge tree a few meters from the riverbank and saw a bedraggled, elderly monkey, crouched alert on the ground and peering toward the group we had just left. Although I came to within just a few feet of him, he looked past me as if I did not exist and kept his eyes on the monkeys in the distance. This monkey was really the worse for wear: he was missing some fingers and his tail tip, and his left hand was completely mangled and inflamed. One of his teeth protruded through his lip. But despite the new injuries and the numerous new age spots and scars, he was still clearly identifiable as Curmudgeon.

I wanted to scream and jump up and down, but of course this would have revealed his location to the rest of the group, so I restrained myself. Todd had guessed who this might be and patiently listened to my excited, whispered babblings about the greatest monkey in capuchindom. I couldn't take my eyes off Curmudgeon. Although he had lost a lot of weight and obviously lost many a fight, he still had the same intelligent look that was so endearing before. He was totally absorbed in spying on Rambo's group, and his eyes darted from monkey to monkey, following the action. I thought, as I had thought so many times, how much better my understanding of monkey social dynamics would be if I could see the world from Curmudgeon's perspective for a few days. Curmudgeon occasionally shifted positions, in total silence, so as to get a better view of the proceedings on the slope below. His left hand was completely useless, so he walked on just three limbs.

Curmudgeon's stealth worked for only so long, however; eventually he made eye contact with another monkey. He made a little pirouetting gesture, as monkeys do when they are initiating a dance, but the other monkey alarm called and screamed. Curmudgeon began to lose his cool and started limping up the hill to find a new hiding place. He paused for a rest part way up, headflagging to the monkey who had spotted him and then threatening me, as if to say, "Let's show some solidarity against this human!" Some young adult male, either Rambo or Q, began to charge toward Curmudgeon, and he turned tail and ran. It was astounding how quickly he could run without the use of his injured hand, and despite our best efforts to keep him in view, he eluded us and the monkey group by heading straight up a steep, crumbling cliff. Frantic at having lost Curmudgeon so soon after finding him, we cursed our bad luck and searched the area for an hour, to no avail. But at least we knew now that he was alive, and as sneaky as ever.

I was struck once more by the asymmetry of affection in our interactions with the monkeys. Curmudgeon had obviously recognized me, because he let me approach him, without reacting, even in his wounded and agitated state. But whereas this meeting was the highlight of my year in the field thus far, Curmudgeon had barely acknowledged my presence. Of course, this was the ideal I had worked toward: monkeys who were so habituated to me that they would proceed with their daily lives as if I didn't exist. But it was nonetheless strange to care so deeply about an animal who was capable of ignoring me so completely.

For the rest of my time at Lomas that season, I thought about Curmudgeon almost daily. How often did I walk right under him while looking for other monkeys? How often did the study group pass by him without seeing him? It was frustrating to think about all of the near misses I had probably had over the past few months, and to know that I was still missing out on so much of his life. Did he have any friends, or did he spend all of his time in isolation now? It was hard to picture such a gregarious and socially intelligent monkey living in complete isolation. And who had inflicted those wounds on him?

A couple of months later, on June 13, we were following Rambo's group when I heard a long and flamboyant series of brays in the distance. Several monkeys responded with intense vocal threats and alarm calls. Suspecting an intergroup encounter, Julie and I ran in the direction of the calls and found Rambo and Q, stuffing their faces with figs and food calling, in the company of another male. I recorded my first impression of this monkey, describing him as "a big tough-looking male who looks like Mudge in his glory days," and when Julie and I got closer we were delighted to find that it was, indeed, Curmudgeon. He was no longer limping, his finger stubs were nicely healed, and he had put on weight since my last encounter with him. His hair was once again standing on end, and he was exhibiting the same vocal exuberance that had made him so noteworthy as an alpha male.

We marveled at how comfortable Curmudgeon seemed in the presence of Rambo and Q, sharing this food source as if they were long-time acquaintances. Rambo and Q had often wandered off from the group during this field season, and now we wondered whether they had socialized with Mudge on previous occasions as well. It was odd how males who seemed to be mortal enemies when they encountered one another in full view of the rest of the group could engage in relaxed social interactions when there were no females or alpha males watching. Often we had seen males vehemently threaten and chase one another during intergroup encounters, for example, only to sneak back for some friendly interaction once all other members of their groups had left the area. Shortly before nightfall Rambo and Q returned to their group, leaving Mudge on his own. Julie and I waited until he had settled into his sleeping tree so that I could find and follow him the next day.

I arrived at his sleeping site before dawn on the following day. At 5:00 A.M. I heard some lost calls in the distance. Curmudgeon became quite agitated and began to produce alarm calls at a high rate. His long sequences of alarm calls were punctuated by an occasional anguished-sounding lost call that reminded me of the early-morning vocal outbursts he used to have in the month following his downfall from the alpha position in Abby's group. It struck me that being solitary might be even harder for Curmudgeon than it would be for a typical capuchin male, because Curmudgeon had always been so intensely social. Perhaps his "fig party" with Rambo and Q the evening before had reminded him of better times, and made him anxious to renew their acquaintance.

After about fifteen minutes of constant alarm calling, Curmudgeon began to run up and down hills at full tilt. I scrambled after him as fast as I could. He began to add bursts to his alarms and lost calls, and there was a sense of urgency to all of the calls. For forty minutes I crashed through wasp-infested vine tangles after him, determined not to lose him any sooner than I had to. Soon we reached forest that I had never seen before. Finally he heard some lost calls in response, from multiple individuals. Instead of responding with more lost calls or running toward the calls, he suddenly fell silent. He shifted direction, so as to move away from the callers. I cursed my luck as Curmudgeon plunged over a cliff and vanished. It seemed that he was determined to avoid the monkeys he had just heard.

I stood there on the cliff for a while, debating what to do. A taxi was due at the farm in an hour's time, to take me into town to buy the week's food for the camp. It was tempting to stay in the forest with Curmudgeon and hope that Joe would go into town on his own to buy the food, but I knew if I didn't return as scheduled he would worry that I had had an accident. Reluctantly, I decided not to follow Curmudgeon. I was curious to see whom he had so assiduously avoided, and so I decided to find those monkeys and then head back to camp after I had identified them.

Just a minute or two after Curmudgeon disappeared over the cliff, the first of the calling males came into view. I recognized him as Porthos, a young male I had seen often lately in an all-male group. Porthos glared at me and uttered a single, soft alarm call. Seconds later his companions, Athos and Aramis, appeared, and they all began to glower and alarm call at me. I could see why Curmudgeon might be frightened of them. They were all young full-grown adults, perhaps ten to twelve years of age: hormone-crazed, in the prime of their life, migrating for the first time, and willing to take any risk. Not one of them had a scar, but they were bouncy and tough looking. This three-male band had been cruising the forest for the past month, turning up in the home ranges of both Abby's group and Rambo's group. They seemed fairly tightly bonded (most likely they were agemates from the same natal group and had played together for years), and they exhibited very little fear of humans.

Did Curmudgeon know these males? Could they be the ones who had injured him two months previously? Just two weeks after this incident the trio successfully invaded Rambo's group, and Q and Feo disappeared from the group shortly thereafter. Porthos would later become one of the most formidable fighters in the history of the field site, participating in the killings of multiple males, and both Porthos and Aramis eventually became alpha males. So Curmudgeon had been wise to avoid them.

Fortunately I did not have long to wait before I saw Curmudgeon again. Two days later, I was once again searching for monkeys and heard an exchange of lost calls from an area we call Heat Stroke Valley. I ran toward the vocalizations and was delighted to find that Curmudgeon was one of the callers. Within ten meters of him was another monkey whom I did not immediately recognize. He seemed perfectly tame, walking on the ground within a meter of me, and he had several scars on his face that did not match any I had ever seen. I radioed Julie to come join me, and also field assistants Kathryn Atkins and Alicia Steele. Julie and Kathryn followed Curmudgeon while I followed the other male, and Alicia, who is an excellent artist, began to sketch the mysterious male from all angles. (As on all days when we encountered Curmudgeon during this field season, we had neglected to bring a camera.)

After staring at the male for nearly an hour, trying to mentally erase the scars and picture him as he would have looked a couple of years earlier, it finally dawned on me who it was: Quizzler, a young male who had been born in Abby's group and who had been a great admirer of Curmudgeon's during his phase as alpha. Quizzler (we later found out he was Curmudgeon's son) had stopped supporting Curmudgeon when Paul Bunyan took over, thereby earning the nickname Quisling, after the Norwegian Nazi collaborator. Julie had reported the previous year that Quizzler was actually alpha male of Abby's group, right after Curmudgeon, Paul Bunyan, and Ichabod vanished. But he had emigrated before I returned to begin this field season, and I had not seen him since he had

gained weight and earned his scars. He still had the same wrinkly forehead and quizzical expression, though, that he had had as a juvenile. Joe
had seen Quizzler earlier in the field season, with his brother Kola in an
all-male group with a third male we did not know, about three kilometers from this area.

Quizzler and Curmudgeon contentedly foraged on nancite fruit and
insects, food calling and lost calling to each other at regular intervals
and maintaining visual contact. Both males were injured—Curmudgeon
was limping slightly, and Quizzler wasn't using his left foot at all. Curmudgeon's hair stood on end more than Quizzler's, but this was the only
clue as to which of the two was dominant. After three hours of foraging,
peeping and lost calling, and watching each other, they had their first direct social interaction. Curmudgeon, who had been foraging tranquilly
next to us for hours, approached Quizzler to five body lengths and suddenly threatened us. Quizzler did likewise, and then rushed over to
pounce on Curmudgeon in an overlord position against us. After breaking off their threat, they foraged on wasp nests within a length of each
other. Quizzler began to look slightly nervous about being in such close
proximity to Curmudgeon and left him, but Curmudgeon followed and
they headed for the same fig tree that Curmudgeon had shared with
Rambo and Q three days earlier. Curmudgeon moved to within contact
of Quizzler, and they rested together briefly before again foraging on figs.

Over the next hour they came into physical contact a few more times,
sometimes presenting to each other for grooming before resuming their
fig foraging. They food called to one another constantly as they foraged.
After leaving the fig tree, they prepared for a longer rest. Quizzler peeped
to Curmudgeon and presented for grooming. Mudge moved closer and
also presented for grooming. Both of them peeped, scratched themselves, and presented for grooming again. Although neither groomed the
other, they finally settled down with Quizzler's tail draped over Curmudgeon's body, looking perfectly relaxed. Later in the afternoon they had
one more fig feast (during which both provided a continuous food-call
commentary), and they chased an adult coati from the tree. We followed
them till 5:00 P.M., when it began raining too hard to see anything.


The bad weather continued, and we searched the forest for the next
three days without finding a single monkey. On the third search day, my

route intersected with Julie's at Bovine Heights, a part of the forest that afforded a particularly nice view. We were frankly getting pretty exasperated about accomplishing nothing, so we decided to sit down and have a chat about her research. After talking for about an hour, I glanced up and was startled to see a monkey napping right above us. Feeling quite silly and incompetent for not having noticed this earlier, we sprang to our feet to identify him. "It's Curmudgeon!" I called out excitedly. He was looking exceptionally fluffy and comfortable, and completely uninterested in our presence. Julie and I admired him from all angles and reminisced about old times. This time we had a camera and were able to take several photos, to his annoyance. Because I was afraid I would not see him again and wanted to have a good photo-ID record in case future field assistants encountered him, I made an exception to my usual rule and used a flash for some of the pictures. This irked him, and he kept repositioning himself to avoid looking toward us. We stayed with Curmudgeon for two and a half hours, during which time he did nothing but eat figs and groom himself a bit. His only companion was a male coati, who was also eating figs.

This turned out to be the last time any of us would ever see Curmudgeon; had I known that, I would have stayed with him longer. This is one of the most frustrating aspects of this sort of research: the individual animals you come to know so well typically vanish without warning, so that there is no good-bye, no closure. Even today, fifteen years after I began my project, Curmudgeon remains one of my favorite monkeys, largely because he taught me so much about capuchin life. I had the pleasure of knowing him as a flamboyant long-term alpha male, a "retired" alpha playing the role of subordinate in the group where he had fathered so many offspring, and as a "lone" male.

Curmudgeon's solitary phase fascinated me because it made it clear that lone males are actually part of quite a large social network. In the brief glimpses we had of his bachelor life, he had proved to know a good many different males, both from social groups and from all-male bands. It seemed that this final phase of his life was every bit as socially challenging as life in a group. He had to be constantly vigilant, assessing who was in the vicinity and whether they were dangerous or not. Given the fickle allegiance and impetuousness of capuchin males generally, it seemed safe to assume that a given male would respond to Curmudgeon differently depending on the immediate social context—for example,

Rambo and Q were happy to associate peacefully with him when they were on their own, but if their social group was nearby they behaved aggressively toward Mudge.

Lone males must have to collect and process information quickly to avoid being attacked. Being a lone male was obviously dangerous, judging by the number of new wounds Curmudgeon had received. The process of making new and trustworthy allies seemed difficult and perilous. I longed to know more about the life of migrant males, but finding and following solitary, habituated (and therefore silent and unperturbed) males in the vast forest was a truly daunting challenge.

Between earning a doctorate and acquiring a tenure-track job, an academic's career resembles the migration phase of a male capuchin monkey's life. In both situations, many formidable competitors vie for scarce slots in stable social groups. Fear of rejection and anxiety about the future are the prevailing emotions. In 1993, after our extended "juvenile" phase at the University of Michigan, Joe and I were bursting with new ideas for research projects and eager to migrate to a comparable institution where we would find a lively community of like-minded scientists and opportunities for additional fieldwork.

Unfortunately, secure jobs in biological anthropology are few and far between, and we quickly realized that we would have to fund our own research during this lean period. We were no longer eligible for graduate-student research funding and not yet eligible for standard research grants, which must be administered by a university and are therefore restricted to faculty members. Our situation was made both more bearable and more difficult by the fact that Joe and I were "migration alliance partners" as we moved into this phase of life. We took turns giving each other moral and financial support, and in helping each other with job applications and write-ups of our data.

Joe is five years older than I am, and he got his PhD four years before I did, so he was the first to enter the job market. We were keenly aware that I was an impediment to his chances of getting a good job. University administrators dislike having to deal with the "two-body problem," and Joe's prospects were further complicated by the fact that he was married to a graduate student who did exactly the same sort of research he did. It was obvious that once I finished my degree I would want a job too, and no one wanted two additional primatologists in the same department—indeed some anthropologists told Joe during job interviews that, in their view, even one more primatologist was one too many. As the rejection letters rolled in and I came closer to finishing my dissertation, we began to consider increasingly drastic and creative ways to continue with our research, such as applying to graduate school all over again, perhaps in psychology or zoology.

Having already applied for, received, and exhausted all possible sources of graduate-student funding during my dissertation fieldwork, we now began to fund the capuchin research entirely out of our pockets, using savings from our teaching-assistant stipends and the small salaries we earned from teaching an occasional course as a substitute for a regular faculty member who was on leave or sabbatical. Whenever we had a break between teaching jobs, we flew to Costa Rica to take a census of the monkeys at Lomas Barbudal, conduct pilot studies for projects we hoped to launch later, and collect fecal samples for eventual genetic and hormonal analysis.

Without outside funding, we could not afford to rent a house or a vehicle. We persuaded the Rosales family to let us camp on their twenty-five-acre rice farm, which was conveniently located right in the middle of Abby's group's home range. We had become quite good friends with this generous and good-humored family, and our arrangement proved to be a socially pleasant, mutually beneficial one. They let us use their stove in the evening to cook our dinner, and in exchange we provided them with an occasional meal and constant entertainment.

The Rosaleses were endlessly amused by our strange foreign ways, and completely mystified by our obsession with monkey feces. We thus served the function of court jester, or perhaps television set, breaking up the monotony of life on the farm with our comic blunders. Every time we cooked dinner or did our laundry, Rosales women stood around us offering suggestions for improving our technique. They laughed till tears came to their eyes at our attempts to adapt our American recipes and domestic routines to life on their farm. They disapproved particularly strongly of our improvised Thai-style peanut sauce, and told even passersby that their silly gringo guests habitually ruined chicken by smearing it with peanut butter, lime juice, ginger (which they use only for medicinal purposes), and spices. The bemused women shook their heads in dismay as I clumsily made amoeba-shaped tortillas and hung my grayish laundry on the line to dry, saying it was lucky I already had a husband, because I sure couldn't marry a Tico if I couldn't make round tortillas and produce bright white laundry. ("Tico" is a polite and affectionate synonym for Costa Rican.) In fact, they sometimes covertly rewashed my laundry so that passing neighbors wouldn't see the clothesline and think that their family was so inept in the womanly arts.

It made no sense to them that I cared more about amassing a large quantity of monkey feces than about being *bien presentado* (well groomed), but they were very tolerant of my eccentricity. After a bit of pleading, I eventually talked them into allowing me to store some of my more valuable fecal samples in their kerosene-powered freezer. Samples destined for genetic analysis were stored in ethanol or silica gel, whereas those intended for hormonal analysis were painstakingly dried in a cardboard solar oven, in little tins. Whenever we were anxiously fussing over our solar poop-drying oven, with its tiny little "pie tins" full of monkey diarrhea, we could count on certain members of the Rosales family (who clearly believed that if a joke was funny once, it was funny a thousand times) to stop by to inquire, *"Siempre la misma receta?"* (Are you making the same recipe again?)

Our two-person dome tent was also a great source of amusement to our hosts. They could not believe that anyone would sleep outdoors with nothing but a flimsy piece of fabric between them and the outside world. At first we could not see what the problem was. Some of the women muttered about malevolent ghosts ("But couldn't ghosts go through wooden walls just as easily as cloth ones?" we asked them), and another family member said, "Why, someone could hack through that with one slash of a machete!" ("But why would they want to?" we asked.)

The real menaces, as it turned out, were the trees and the livestock. After some frightening incidents with falling branches and stampeding cattle (driven to within a foot or two of our tent by a drunken cowboy at midnight), we moved the tent so that it was protected by the awning of the fertilizer shed. As it turned out, this was the most popular spot on the farm, because the tin awning offered partial protection from the heavy rains. The smaller farm animals (pigs, dogs, cats, and chickens) deeply resented our intrusion into this bit of premium real estate. We were awakened several times a night by roosters crowing six inches from our ears (note to city dwellers: they don't crow only at dawn) and by loud fights as animals jockeyed for the best positions. The pigs and dogs leaned against the tent, craving the warmth of our bodies during the cool predawn hours, so that they were virtually in bed with us. Two of the pigs were disconcertingly large and would practically flatten the tent when they flopped down against it. We frequently had to poke and punch them through the tent wall to get enough space to sleep comfortably. It was crowded enough in there with the two of us and all our field equipment, clothes, toiletries, and hundreds of monkey fecal samples; we needed all the space we could get. (It didn't smell so great either.)

By far the most disgusting aspect of our living arrangement with the animals was the fact that the pigs preferred to defecate on the path from our tent to the distant outhouse. Answering nature's call in the middle of the night, our way guided by a flashlight with perpetually weak batteries, we often stumbled right into a steaming mound of fresh pig poop and felt it squelch between the toes of our sandaled feet. Without running water or electricity, there was little that could be done to rectify this sad state of affairs, and we'd have to crawl back into bed, pig poop and all, and wait for the sun to rise. We tried to alleviate the smell by hanging four Ty-D-Bowl toilet sanitizers from the ceiling of the tent, but this only enhanced the tent's latrine-like ambience and made permanent pink and green stains on the bed sheets as the humidity condensed on the sanitizers and dripped down. Most couples have their salad days; we had our pig-poop days.

It was sometimes difficult to explain to our friends and relatives why monkey poop is of sufficient scientific value to warrant such a lifestyle. In fact, the monkeys' waste is chock-full of exciting information that cannot be obtained simply by watching the animals. Obviously it contains remains of the foods they ate, including bits of insects that were unidentifiable before consumption because they disappeared into the monkeys' mouths a fraction of a second after capture. By examining the state of defecated seeds, and attempting to germinate them, researchers can gain insight into the nature of the digestive process and the co-evolution between monkeys and the plants whose seeds they disperse. However, we were primarily interested in the social information that the feces could yield.

Feces contain hormones (both reproductive hormones, such as estrogens, and the stress hormone cortisol), and we knew that these hormones could give us insight into the animals' reproductive states and stress levels, thereby telling us all sorts of interesting things about reproductive strategies, and in particular about reproductive suppression—the use of harassment to induce so much stress in same-sex companions that their reproductive function is impaired. We conducted a pilot study on this topic but the project met a dismal end when our laboratory collaborator left science altogether and also lost our samples.

For us, by far the most exciting potential use of feces is as a source of DNA. Fecal waste contains cells sloughed off in an animal's digestive tract, and during the early 1990s geneticists were developing techniques with which to extract and amplify DNA from these cells and to distinguish the individual's own DNA from potential contaminants.[4] These advances meant that, for the first time, it was possible to determine wild animals' genetic "fingerprints" without trapping, darting, or otherwise disturbing them. For both ethical and practical reasons, we were reluctant to impose any form of stress on the monkeys. Researchers at others sites had told us that white-faced capuchin troops sometimes dehabituated after one of their members was anesthetized and temporarily captured.

Genetic information would advance our understanding of capuchin social behavior on two fronts. First, we would be able to determine which males had fathered which infants. A long-running debate in primatology concerns whether dominant males are more reproductively successful than subordinates. Since animals strive for dominance, and since status competition entails costs such as increased risk of injury, Darwinian theory predicts that high rank pays off in the form of greater baby production—the ultimate currency in the marketplace of evolution. Dominant males are thus predicted to exclude subordinates from mating opportunities and to be more attractive to females. However, theory can only pose questions, it can't answer them. Skeptics rightly demanded to see the data supporting this hypothesis.

As confirming evidence rolled in from several species (red howler monkeys, long-tailed macaques, and yellow baboons, for example), the debate itself evolved into a set of more subtle and interesting questions.[5,6,7,8,9] If dominant males outreproduce subordinates only under certain conditions, what are those conditions? When all the females of a group are sexually receptive simultaneously, even the most despotic alpha male might be unable to keep an eye on them all, with the result that subordinates will father some offspring. Male reproductive success might be more balanced if subordinates formed aggressive coalitions against dominants, blunting their competitive advantage. Finally, females themselves are not mere passive resources but active agents with their own evolved interests, which are sometimes surprising.[10] One of Joe's most striking results from his study of the Cayo Santiago rhesus macaques was that sexually receptive females actually maintained spatial proximity more actively to subordinate males than to dominants.[11] We were eager to discover what our capuchins could contribute to the resolution of these questions.

Our second planned use of the monkeys' DNA was to determine, at least roughly, the genetic relationships among all members of each troop, including the adults. Possibly the late twentieth century's greatest breakthrough in understanding animal social behavior occurred in 1964 when William D. Hamilton showed mathematically that genetic kinship is central to understanding the phenomenon of altruism, in which an individual does something that increases another's reproductive success at some cost to its own.[12] Because closer kin share more genes inherited from recent common ancestors, a gene for altruism can flourish so long as it programs its carriers to dispense altruism discriminatingly, favoring closer relatives over distant ones, and kin over nonkin. For example, because full siblings share half their genes by recent descent, altruism toward full siblings will be favored, so long as the reproductive cost to the altruist is less than half the reproductive benefit to the beneficiary.

Hamilton's theory, known as the kin-selection or inclusive-fitness theory, had helped solve numerous puzzles, ranging from the sterile castes of social insects to the "helpers at the nest/den" phenomenon found in several birds and carnivores, in which young adults remain with their parents and help rear their younger siblings rather than leaving home to breed. Several capuchin behaviors appeared altruistic, including defending companions from aggression by higher-ranking monkeys and nursing other females' infants. We wanted to know if genetic kinship helped explain which monkeys provided these services to which others.

It is funny how humans can be trained to appreciate substances that are initially repugnant. No sane person would gain any aesthetic pleasure from his initial exposure to capuchin feces. Whereas a folivore's poop is fairly innocuous in form and odor, an omnivore's feces is invariably foul. Capuchins produce a runny, typically black, sticky splatter of glutinous slime with an odor so distinctive and potent that it can be used to help track the animals. But we have come to associate this smell with the presence of monkeys and with the production of genetic samples. So whereas normal people would respond to the odor by hastily beating a retreat, we and our dedicated field assistants run joyfully toward it, whooping with glee at the sweet (to us) smell of success and data, just as Pavlov's dogs were conditioned to be aroused to a state of eager anticipation at the sound of his bell signaling feeding time. It is not uncommon for us to stand around simply admiring a particularly fine (i.e., easy to collect) specimen, rueful that we do not know which monkey it came out of and therefore cannot use it. Our lab collaborators often have to curb our enthusiasm, reminding us not to get too greedy and put more feces in a collection tube than is appropriate for the amount of silica or solvent that will be added. Often field assistants just can't bear to leave behind any part of a specimen that comes from a favorite animal, or from one they have been following for several hours just for the purpose of collecting the poop.

For many years we hauled our fecal samples from place to place, storing them in five different states, in the freezers of friends, family, and the people subletting our apartment. Every trip through customs was fraught with tension, as the laws regarding importation of monkey poop are not particularly well defined and we knew we would be at the mercy of the customs staff of the day. I recall one particularly memorable transit day in 1994, when we had samples in film canisters, which had leaked a bit into the Ziploc bags we carried them in. We sheepishly presented the customs agent with a bag of canisters floating in what seemed to be monkey diarrhea, which we anxiously reassured him had been sterilized in ethanol. To my immense relief, he just registered a look of appalled disgust and hastily ushered me through the line. Our experience could have been much worse. One of our colleagues tells a story about losing an entire year's worth of chimpanzee feces when his suitcase was stolen on a train in Africa; his agony was mitigated only by his amusement at picturing the response of the thief who was expecting saleable goods and found the suitcase to be packed with chimp shit.

As the size of our collection grew, and the years ticked by without any progress being made on the lab work, it became increasingly difficult to convince our freezer "hosts" of the scientific worth of these samples (that is, to persuade them that the vials of monkey excrement deserved prime space in the coldest part of the freezer, when they would rather store their ice cream there). We often wondered whether samples were removed during our field stays and hastily replaced in the freezer just before we returned.

After three graduate students from different universities had half-heartedly attempted to analyze our samples and then abandoned the project before making any significant progress, in 2001 we finally established a collaboration with Linda Vigilant, widely acknowledged to be the world's expert at extracting usable DNA from primate feces. She agreed to analyze the samples in her lab at the Max Planck Institute for Evolutionary Anthropology in Leipzig. We then persuaded a Brazilian graduate student, Laura Muniz, to do her dissertation work on our samples, under Linda's supervision.

At last, a decade after we began collecting fecal samples, we started to get the long-awaited results. To our immense relief, even those leaky film canisters had survived the summers in our sublet freezers and yielded high-quality DNA. Dozens of field assistants, past and present, waited along with us to learn who was the father of whom, and we pestered Laura regularly for news. This baffled her at first, for to her the samples were just test tubes with initials on them, not the keys to information about favorite characters in an ongoing soap opera.

In the great majority of monkey and ape species that live in groups containing several adults of both sexes, males and females often mate with several partners within a short time. Nevertheless, whenever primate mating patterns have been studied in detail, it has become clear that individuals of both sexes prefer some mating prospects to others. Mate choice by males has received relatively little research attention, but some evidence suggests preferences for high-ranking females and fully adult (as distinct from adolescent) females, as would be expected given their higher probabilities of successful reproduction. Female choice has been studied in greater depth and in a larger number of species. It is increasingly clear that male mating success is sometimes just as strongly influenced by female preferences as by male dominance rank.[10,11,13]

What qualities might be desired in a monkey or ape mating partner? Primatologists have suggested many answers but have found only limited support for most of them. For example, Barbara Smuts's groundbreaking 1985 book *Sex and Friendship in Baboons* is still one of the only studies to support the plausible and heartwarming hypothesis that females in promiscuous groups prefer to mate with their "friends": males that had formed friendly, protective relationships with the choosing females and their infants (even those sired by a different male) before the female became sexually receptive.[14] Classic sexual-selection theory tells us that females should prefer males with elaborate ornamental traits (think of the peacock's tail) and showy displays designed to appeal to the opposite sex. Though these attributes may yet prove important in primate mate choice, very little is now known about their effects.

Two ideas about primate female mate choice have received strong support. One, proposed originally by Sarah Hrdy some thirty years ago, is that females should prefer unfamiliar males to familiar partners, because while any new male is a potential perpetrator of infanticide (see Chapter 8), even a small probability of having sired an infant will deter a male from killing it.[15] The second hypothesis is that both males and females will avoid mating with close kin, to reduce the possibility of genetic defects in their offspring, but females will be particularly averse to inbreeding.[16] The difference is expected because females, with their greater investment of time and energy in each offspring, have more to lose than males from conceiving a defective infant. As in most mammals and birds, "opportunities" for inbreeding are rare in primates, because usually all or almost all adolescents of one or both sexes leave the group of their birth to seek their fortune elsewhere.[17] This means that, as adults, close relatives of the opposite sex don't see each other except perhaps occasionally from opposing sides of an intergroup battle. In fact, various lines of evidence suggest that sex-biased dispersal, as this pattern is called, has evolved specifically to prevent inbreeding. In most mammal species it is the male adolescents who are struck by wanderlust (in birds, it's usually the females). But Tim Clutton-Brock has shown that in mammal species in which adult breeding males typically stay in social groups for so long that their daughters have reached sexual maturity, females are the dispersing sex.[18] This makes sense if the undesirable possibility of father-daughter inbreeding has favored females who leave home and family behind.

Natural selection has apparently installed in primates at least one fail-safe mechanism to prevent close inbreeding even when coresidence makes it a possibility. The hypothesis that humans are not sexually attracted to those with whom they were close childhood companions, proposed by Edward Westermarck more than one hundred years ago, has in the past two decades received convincing support from several sources.[19] At some times and places, unrelated children have been raised together and either not discouraged from marrying, as on Israeli kibbutzim, or actually required to marry, as in a Taiwanese practice known as "minor marriage." Such children, on reaching adolescence, felt little or no attraction toward their pseudo-siblings.[20,21] Marriages between kibbutz agemates occurred far less than expected by chance, and minor marriages had lower fertility and were more likely to end in divorce than other arranged marriages. In two experimental studies of American college students, those who had grown up with opposite-sex siblings were more likely than those who hadn't to endorse harsh punishment for a fictitious couple engaging in consensual sibling inbreeding.[22,23] It appears that having a close companion of the opposite sex in early childhood activates a specific psychological mechanism producing strong aversion to brother-sister sexual relations.

Does nonhuman primate psychology also include sexual aversion toward close childhood companions? A wealth of data from many species indicate that in the unusual cases in which some individuals of the typically dispersing sex stay and try to breed in their birth group, they mate far less than expected by chance with their maternal relatives.[24] The few copulations between close kin that do occur tend to result from males coercing their sisters into mating. Furthermore, Joe and I showed for rhesus macaques, and Joseph Soltis and his colleagues showed for Japanese macaques, that female macaques in estrus were more responsible than their adult male relatives for keeping a large distance between themselves and the males. These findings confirm predictions from the theory that females are more averse to inbreeding than males.[25,26] Perhaps the most striking evidence for inbreeding avoidance in monkeys comes from observations of female homosexual interactions in Japanese macaques. These interactions are quite frequent during the mating season, although the participating females also copulate with males. Females almost never choose related female partners, even though they prefer their kin for nonsexual friendly acts, such as grooming and huddling.[27] Nevertheless, it remains a matter of debate whether these results support only Westermarck's "familiarity does not breed" hypothesis, or whether close kin recognize each other more directly by checking (not necessarily consciously, of course) whether their own inherited characteristics match those of prospective mates. In most primates, maternal kin, unlike paternal kin, usually grow up as close playmates, because of primates' close and long-lasting relationships with their mothers. The key test cases for deciding whether kin recognition could operate via "phenotype matching," as distinct from familiarity, are cases in which close kin are not close childhood companions. Hints of phenotype matching in primates come from recent observations that female rhesus macaques and savanna baboons form closer bonds with *paternal* half-sisters (females with the same father but a different mother) than with unrelated females.[28] Additional results from the same baboon population hint that opposite-sex paternal half-siblings are somewhat averse to mating.[29]

In addition to brother-sister inbreeding, there are two possible forms of inbreeding among primary kin: mother-son and father-daughter. Assuming that these kin pairs have not been separated by kin-biased dispersal, the modern version of the Westermarck hypothesis predicts that father-daughter inbreeding will be the most common of the three forms, and data on humans bears this out. The theoretical reasoning is three-fold.[30] First, when only one of a pair of relatives is a child during their initial relationship, the adult of the pair will develop a weaker aversion to sexual contact than the child. Second, for reasons already described, males are less averse to incest than females. Third, fathers are larger and stronger than their daughters and can therefore pressure them into unwanted sexual interactions.

How frequent is father-daughter inbreeding in nonhuman primates when, as occasionally happens, fathers and their adult daughters reside in the same social group? Among the mountain gorillas made famous by Dian Fossey's research, females reaching maturity in the same group as their presumed father—that is, a male who had held alpha rank since the female's birth—were almost never observed copulating with them.[31] A study of captive rhesus monkeys, incorporating genetic testing, showed that although paternal half-siblings, paternal uncle-niece pairs, and other classes of paternal kin conceived offspring together no less frequently than expected by chance, father-daughter pairs did avoid mating.[32] But Jutta Kuester and Andreas Paul found no signs of father-daughter inbreeding avoidance in captive Barbary macaques, a species that (like rhesus macaques) lives in large promiscuous groups in which males show no inclination to associate preferentially with their own offspring.[33] However, matings were very rare between maternal kin as well as in unrelated male-female pairs that had interacted frequently when the female was an infant. Kuester and Paul argue that these results support only the Westermarck mechanism and cast doubt on the possibility that primates recognize kin by direct assessment of inherited similarities. Of course, captive monkeys lack wild monkeys' opportunities to disperse to new groups when suitable mates are scarce. No researchers had used genetic data to see whether nonhuman primates avoid father-daughter inbreeding under natural conditions.

In 2004 the genetic results on our population of capuchins gave us exactly this opportunity.[34] The first surprise in the data Laura and Linda produced was the discovery that some alpha males had a reproductive monopoly that lasted many years. We had always expected the alpha male to have a reproductive advantage over the other males in a group, but we were stunned to find out how big the advantage was. The alpha male sired 80 percent, 95 percent, and 100 percent of the offspring of his nondaughters in the three groups for which long tenures were observed. Curmudgeon sired at least twelve offspring. If our guesses about the ages of the juveniles present in 1990 are accurate, then Curmudgeon was alpha male for about eight years before his downfall in 1992, and the only infant not sired by him was Lizano, the son of his daughter Nanny. Similar reproductive monopolies were achieved by Pablo and Fonz, who have held alpha male positions in Rambo's group and Pelon group for about fifteen or sixteen years each. Pablo's sons are alpha males in three adjacent groups, and two neighboring groups are populated by his daughters, so his reproductive influence extends well outside his own group. In all three cases of long-term alpha males, there was father-daughter inbreeding avoidance, such that the alpha male did not sire the offspring of his daughters (with one exception out of seventeen cases in which a female conceived an offspring while her alpha male father was still in her group). Thus subordinate males could achieve a fair measure of reproductive success in groups in which the alpha male had been in place for many years and had multiple reproductively mature daughters.

The degree of monopolization by alpha males was extraordinary for many reasons. We frequently observed copulations between subordinate males and adult females. The alpha male typically tolerated these copulations, even when they occurred in full view of him, perhaps because the females in these copulations were typically pregnant or lactating and therefore incapable of conceiving. Nevertheless, this tolerance seemed to contradict the genetic results implying extreme reproductive competition. How did the alpha male achieve such a monopoly? Mate guarding would be extremely difficult in this species, because the dense foliage and wide group spread during foraging make it easy for females to be out of view of the alpha male for long stretches of time. We had very rarely observed anything that looked like "consort behavior," and to the extent that we had, it was usually the female shadowing the male rather than vice versa. Capuchin males seem more interested in food than in sex in many cases, and we have even seen alpha males slap females who are pestering them for sex. One frustrated adolescent female, desperate for attention from the alpha male, bit him on the tail and pushed him out of the tree when he persisted in eating instead of responding positively to her advances.

The strong sexual assertiveness of females, combined with the difficulty of mate guarding, make it likely that females actively choose the alpha male. Surely subordinate males would be able to achieve sneaked copulations if they had any cooperation at all from the females. But why do the females prefer the alpha male so much? Charles Janson has proposed for the closely related *Cebus apella* that females prefer the alpha male because he controls access to monopolizable resources, like palm fruits.[35] The alpha then tolerates cofeeding in these palm-fruit clusters by his offspring and mates. However, the foods eaten by white-faced capuchins tend to be far more dispersed than palm-fruit clusters, and the alpha males cannot wield so much control over access to them. Therefore a strong female preference for the alpha male would be a bit more puzzling in the case of our capuchins.

For males such as Curmudgeon, Pablo, and Fonz to sire such a large proportion of their groups' offspring, most other males must end their reproductive career with no offspring. Given the intense competition to be a capuchin alpha male, it is remarkable that some males manage to hold on to their alpha rank for more than fifteen years; such long tenures are highly atypical for primates generally. In some of the better-known species, such as baboons, the typical pattern is for a young male in his prime to barge his way to the top of the hierarchy and then stay there just a few short months or years before being toppled by a young successor. How do the capuchin alpha males manage to stay on top so long? Fonz is a burly male even in his old age, but the same is not true of Pablo, who looks rather slender and frail compared with the other males he regularly encounters. Clearly fighting is not everything. Males such as Pablo and Curmudgeon seem to owe their success in part to charisma and social skills. All three of these long-term alpha males seem to be particularly good at avoiding escalated fights, both within and between groups, leaving the dirty work to their allies. It appears likely that natural selection has favored males who have the psychological poise to stay calm under attack, the excellent memory and social intelligence necessary for noticing and remembering the qualities of other group members' relationships, and the social skill to dole out just enough favors to their male allies to maintain their loyalty. In the next two chapters we will examine the social challenges faced by males when migrating, and the supreme importance of having dependable male allies.

One of the most important lessons I have learned in the past fifteen years about studying primate behavioral ecology is that if you want to have a simple story about what the animals are doing, you shouldn't study them for very long! Like most primatologists at the start of their work with a relatively unknown species, I based my early publications on just a couple of years of data from a single social group. My first impressions of capuchin society were of tightly female-bonded groups with very stable membership, low mortality, and relatively amiable relationships among all group members. Now, having followed the monkeys for so many years, I look back on that period of Curmudgeon's reign as the Golden Age of Abby's group. During the first five years of my study, there was only one case of immigration into Abby's group, and there were no emigrations or deaths, aside from the deaths of two very young infants. Besides Curmudgeon's fall from power and the ascent of adolescent Diablita up the female hierarchy, there were no rank reversals, either. Life was relatively tranquil, and I thought capuchins were fairly sweet-tempered toward members of their own species, at least in comparison with the rhesus monkeys Joe and I had studied previously.

Knowing what I now know about capuchin social life, I certainly do not call their social life tranquil (though it has its calm periods, during the reigns of long-term alpha males). Comigrating pairs and trios of males can initiate a domino effect of spreading social chaos. As these co-alitions invade neighboring groups, males from those groups are often evicted, and then they try to invade other groups. Bands of males start roaming the forest and visiting many neighboring groups, looking for migration opportunities, and the society begins to acquire a fission-fusion organization that is essentially absent during more stable periods.[1]

Linda Fedigan, who had been conducting demographic research on capuchins in Santa Rosa National Park since 1986, had told me about observing many such migration waves, but as far as I could tell no such phenomenon took place at Lomas Barbudal in my first six years of research. Although I was loathe to see my beloved study subjects undergo any stress, I was also curious about how the adolescent males in Abby's group would go about forming new relationships in neighboring groups. Because it would be too difficult to document such a process in an unhabituated group, Julie Gros-Louis, Joe, and I decided we would have to habituate more groups. We started with Rambo's group, the long-time neighbors—and enemies—of Abby's group.

As described in Chapter 7, Julie found that major changes had occurred in Abby's group between the summers of 1995 and 1996: all of the adult males had vanished, and the group was "ruled" briefly by Quizzler, a young natal male. Ichabod returned and took over as alpha male in December 1996, but he never earned the loyalty from females and juveniles that other alphas had enjoyed. Nevertheless, we were shocked when he was killed by his own groupmates in March 1998, a little over a year into his term as alpha. Guapo and Hongo both served multiple terms as alpha male, and they were joined a few years later by Thornhill, a snaggle-toothed, mottled old male who also served a brief stint as alpha. In 2001 and 2002, six new males immigrated, some staying only for a short period of time or coming and going periodically. With so many males on the scene, their ranks changed more frequently and the social dynamics became difficult for us to follow—we estimated that the alpha position changed hands thirteen times in 2002, rotating among four different individuals. This instability was completely unlike anything I had seen before (or have seen since) in any of my three study groups, each of which had kept the same alpha male for several years running. Two of the migrants (Jackson and Jordan) immigrated together and were obviously very close allies. Jackson became alpha of Abby's group for the first time in June 2002, with Jordan's aid, and he served several stints as alpha male after that. One of Jackson's first acts as alpha male was to kill Eldritch's baby, Einstein, the youngest infant in the group (four months old at the time of the attack). Sadly, this was not to be the last episode of this sort, although, as explained later in this chapter, such killings are not surprising to behavioral biologists.

The next year, 2003, proved to be an even more stressful year for Abby's group than 2002 had been. The four males who were jockeying for the alpha position continued to do so. And to make matters worse, several males from neighboring groups were forming all-male bands and cruising about the forest, looking a place where they could move in and take over. Abby's group was frequently stalked by Heinrich and Punto, two highly formidable young males from Pelon group, our most recently habituated group.

During the first week of April 2003, Abby's group had two particularly nasty intergroup encounters with Rambo's group and somehow became fragmented into multiple subgroups that were scattered over several kilometers and could not reunite. Various males from Rambo's group began stalking some of these subgroups, leaving their own group for days at a time so that there were hardly any males in Rambo's group. Pelon group also encountered a portion of Abby's group and fragmented it further. From that point on, several of the young adult males from Pelon group began stalking other Abby's-group monkeys. It was as if the word was out in Lomas Barbudal that Abby's group was vulnerable and ripe for a takeover. Lone animals from Abby's group were lost and roamed the forest looking for their companions, trying to switch from one subgroup to another.

Slowly the subgroups began to stabilize a little bit. Solo, a young natal male from Abby's group, became leader of one subgroup that consisted mainly of other young natal males. Moth, a burly, greedy male natal to Rambo's group, assumed leadership of another of the Abby's-group subgroups. Several other males from Rambo's group joined him, but they squabbled with one another constantly rather than presenting a united front, and they seemed nervous when they collided with males from Abby's group. All over the forest there was high drama. Three Pelon males found a former Abby's-group male (Took) and beat him up so badly that we were sure he would die (though we turned out to be wrong). These three thugs (Punto, Heinrich, and Alamasy) then teamed up with some of the females from Abby's group. The main portion of Pelon group came upon the little orphan Nobu, our favorite infant from Abby's group, all by himself. They terrorized him and gave him a thorough looking over but finally moved on, leaving him alone in the river, shivering and foaming at the mouth. Meanwhile, another sizable chunk of Abby's group was led by Weasley, one of the males who had been alpha male off and on during the previous year and a half.

The entire month of April was fraught with danger and excitement as monkeys moved from group to group, trying to decide where to cast their lot. Juveniles had to make hard decisions about whether to stay with their mother or their play partners. Many individuals spent long days and nights completely alone while they searched for the group they wanted to join. Some animals disappeared during this period and were never seen again. Males continued to shift alliances on a regular basis, and most of them were wounded many times during this month. We would have needed a team of about forty researchers to document all of the exciting turns of events. During the day, our walkie-talkies were buzzing with constant revelations about the locations and interactions of missing monkeys and frantic calls for additional observers to help locate parties that had just fissioned from their subgroups. Every night at dinner it took us at least thirty minutes just to sort out who had seen which monkeys where, and to formulate the best strategy for data collection on the following day. Often we were forced to work alone instead of in teams, in an effort to keep an eye on all known subgroups.

It was as if the social organization had switched overnight from stable, multimale, female-philopatric groups to a chaotic fission-fusion society. The females were quite pregnant and visibly stressed as they tried to negotiate relationships with all of the new, hot-blooded males who kept bursting into their lives. It became clear to us that the cognitive challenges faced by capuchins in the social domain were even greater than we had previously thought. Each individual monkey had to make educated guesses regarding the locations of those she or he wanted to be with, recall important information about the reputations of the males in each subgroup, and make life-or-death decisions about which group to join.

By the end of April most of the young natal males from Abby's group had left to form their own all-male group, consisting of seven males ranging in age from four to eight years, and they never again joined the other Abby's-group monkeys. They disappeared onto a private ranch, and most of them have not been seen since. Solo, their leader, eventually took over a portion of Chingo's group and was joined by some other natal males from Abby's group; they occasionally received visits from non-natal former residents of Abby's group as well.

In early May, as the birth season was beginning, chaos was still raging in the remainder of Abby's group. Finally all the adult females had reunited in one group. By this point most of the original adult males of Abby's group (Hongo, Weasley, Jackson, Jordan, and Bentley) had returned, and Heinrich, of Pelon group, was stalking the re-formed group. Celeste was the first of the group's females to have her baby (Cricket), on May 11. We watched anxiously to see what would happen, hoping that there would be no more changes in male membership or rank. Having already witnessed three cases of infanticide, the most recent being Jackson's killing of Einstein, we knew that the impending cohort of newborns would be vulnerable to infanticide. We did not yet know who the fathers in these cases were (and we were not sure whether the fathers themselves knew), but regardless, there were likely to be numerous takeover attempts in the next few months by males that had never mated with the infants' mothers. All the potential fathers had had fairly spotty "attendance records" in the past several weeks, which meant that the probable fathers were unlikely to be on the scene every time their progeny needed to be defended. And with all of the males' relationships in constant flux, it seemed unlikely that the current core of resident males would be able to fend off invasions by the migrating males coming from Rambo's group and Pelon group.

Infanticide is hard enough to witness under normal circumstances, and it was going to be all the more stressful for us this time because the primary focus of our research during this field season was infant development. However, the anxiety of waiting for these infants' fates to be revealed was mitigated by the more pleasurable anticipation of finally getting a chance to document the immigration process of habituated monkeys migrating into habituated groups.

There were enough exciting events during this season to fill numerous books, and I cannot recount them all. Instead, I will report the events of one day in Abby's group during this pivotal time, to convey a sense of the social interactions involved.

**May 17, 2003**    It is just before dawn, and I am looking for Abby's group, accompanied by a new, untrained field assistant. The rainy season is just starting, and the dusty, barren, brittle forest of April has already been replaced by masses of thick, green, soggy, resilient vegetation. The air is cool and damp, the ground waterlogged and slippery. We arrive at the group's sleeping site shortly before sunrise and stand expectantly, listening for the monkeys.

I love this time of day, when the night creatures are going to bed and the diurnal ones are starting to arise. With my vision impaired by the dark, and the monkeys asleep, I open my ears and nose to take in everything around me—a luxury I cannot afford during the daytime, when I must focus strictly on the monkeys. The nighttime frog chorus is still sounding, though it is starting to lose momentum. The air is filled with the rich aroma of rotting vegetation and fungus. I am surrounded by a constant pattering sound, like the sound of rainfall but more gentle. I know from experience that this is the sound of hundreds of thousands of caterpillars defecating on the leaves below them, and I feel these tiny pellets showering down on me, too, punctuated by the occasional heavier plop of a caterpillar that has lost its footing. A bat whooshes past my face, barely missing me. The eerie call of the *cuyeo*, a Central American nightjar, sounds just ahead of me, and I hear the flutter of wings as it rises from the ground briefly and settles back down further ahead on the trail. The atmosphere is tense with the sounds of frenzied biological activity, but I still haven't heard any sounds of the particular drama I am here to document. As the sun approaches the horizon, I can see mist swirling around me, and the dense growth surrounding me starts to acquire some color. The mosquitoes locate me and begin to whine all around my head; I swat them with my floppy hat, unintentionally squashing a juicy caterpillar on my neck. A tinamou pierces the dawn with a clear, pure tone and is answered by a faint echo in the distance. About twenty meters away, an agouti gives an alarm call and then bounces away from me. These predawn moments make the chronic sleep deprivation of my job worthwhile: it is such a pleasure to stand in the forest in the dark, completely surrounded by the sounds of thousands of virtually unstudied organisms, trying to imagine what they are all doing. There is no doubt about it: I have a terrific job.

I am jolted back to the task at hand by the sounds of monkeys waking up, and I press through the rain-drenched vegetation to get closer. Someone is shaking branches, already in the mood to cause trouble for someone else. Although it is too dark to see much, I hear alarm calls, sex vocalizations, and finally a long series of lost calls. Most of my days start this way now, as there are so many lost individuals trying to decide where they belong. I can barely make out the copulating couple, just well enough to see that it is two males. Soon there are vocal protests by many, and monkeys are rushing about wheezing, sounding intense vocal threats, and trilling: the sounds of alliances forming and being tested. As more light begins to penetrate through the canopy, I make out the forms of Bentley (a sweet-tempered low-ranking male who has been residing in the group for more than a year), Jordan (the infanticidal Jackson's sidekick), and Weasley (the current alpha male). Then I see Took, a male who had been resident in Abby's group earlier but had been absent from the group for the past four months. This is exciting: I had presumed him to be dead, knowing how badly he had been hurt in the gang attack by three Pelon males a month previously. Took's wounds are healing very nicely. Most likely the lost calls earlier had been Took and the sex vocalizations had been part of his reunion with the group.

Weasley begins threatening off in the distance, and I turn to see who is approaching. To my amazement I see Moth and Tranquilo, young adult natives of Rambo's group, striding toward the group. Moth (named for some white tufts on the back of his head that resemble the white "paloma" moths that plague our laundry) is the burliest male of his age I have known and has been a big favorite of researchers interested in foraging behavior, owing to his extraordinary gluttony. He is normally too interested in eating and napping to investigate other groups, but in the past month he has been more adventurous. Tranquilo has been Moth's main playmate since we have known them, so I am not surprised to find them together—although I'm a bit surprised to see them so far from home, since this portion of Abby's group's range is not in an area that overlaps with other groups' ranges. Both Moth and Tranquilo are swaggering toward Abby's group, glowering and grinding their teeth together at surprisingly high volume. I have never seen Moth so agitated. He hops up and down and grunts, and Bentley, Jordan, and Weasley lunge at him. Moth and Tranquilo utter intense vocal threats as they jointly chase the Abby's-group males into a dense bamboo thicket. When they emerge, Tranquilo is nursing an injured hand.

The females from Abby's group draw closer together, and the fighting starts up again. Moth and Tranquilo resume their loud tooth grinding, and Diablita and Vandal (the alpha female and second-ranked female) and Weasley scream at them. Jackson has arrived, and he, Jordan, Weasley, and Bentley are absolutely livid, directing intense vocal threats and screams at the intruders. There is a somewhat pathetic, shrill tone to Weasley's voice as he squawks and squeals at Moth; he seems seriously rattled by this challenge. Moth and Tranquilo slowly and deliberately advance on Weasley, making duck faces at him. Weasley responds with pirouettes and sex squeaks, but then Moth and Tranquilo resume their hostile tooth grinding. Weasley rears up on his legs and spins about as he courts his enemies, who continue to move closer. Weasley retreats a bit and screams some more. As Jordan and Hongo approach, Weasley appears to gain confidence and breaks branches. Hongo starts tooth grinding too, and suddenly all hell breaks loose and there are monkeys flying everywhere. Abby's group seems to be winning for a moment, but it is hard to tell what is going on as monkeys run circles around me in the dense undergrowth, chasing and biting one another. When they move into a somewhat clearer area, I see that Weasley has just gotten a huge gash in his neck, which is bleeding profusely. Most likely it was inflicted by Moth, who has a new wound on his wrist. Moth isn't showing the slightest sign of giving up. He and Tranquilo charge toward Weasley, who turns tail and flees.

At this moment Celeste appears with her newborn baby Cricket on her back and begins to threaten Moth and Tranquilo at the top of her lungs. Vandal joins her in threatening. I am tense, expecting a sticky end for Cricket, and then am surprised to see that the would-be immigrants have far more respect for these angry females than they do for the males of Abby's group. They nervously retreat and watch from a distance. Celeste's juvenile son joins in threatening them, as does Jackson. Vandal storms up to within two feet of Moth and screams at him, causing him to bounce away. Diablita joins Vandal in threatening when

Moth shakes a branch. Abby's group surrounds Moth and screams at him.

Moth initiates a sex dance with Jackson, and then three other monkeys dance as well. Vandal lunges and screams at Moth, and Tranquilo loses his nerve and flees. Moth also retreats as Jackson, Diablita, and Hongo pursue the two intruders, though he does so more calmly than Tranquilo did. Then Moth, ever the glutton, spies a wasp nest and breaks off hostilities to devour it calmly while surveying his opponents. As the vocal protests against their presence again increase in volume, Tranquilo stands bipedally to get a better look and Moth starts to look uncomfortable. Jester, a normally quiet and peripheral female who tends to mind her own business so long as her family is not being directly threatened, has joined the other females and juveniles in threatening Moth. Once more, Moth and Tranquilo seem more frightened by the threats of these females (unknown to them) than by the attacks of the males, whom they have been combating in intergroup encounters for the past several years. Their eyes dart back and forth, taking in the females' interactions apprehensively. Moth dances briefly with Jackson and then begins to stalk purposefully toward the females once more. Jester walks toward Moth and threatens him again before losing her cool and backing away a bit. The rain starts to pour down in torrents, and both Jester and Vandal begin to threaten Moth more vigorously.

Moth startles me by trying to slap me. He has never done this before; it appears that he is trying to incorporate me into his display to impress the females. His hair stands on end and he circles me, panting heavily and threatening me. He makes some vicious-sounding growly noises I have never heard before. He looks like he really means business, and I get nervous, turning around as he circles me, so that I can see what he is up to. Tranquilo rushes over to join Moth in an overlord against me, and they bounce at me. Normally I am not unnerved by capuchins' threats toward me—they are usually just bluffing. But when males are really pumped up, as they are during takeovers, they become impulsively violent and sometimes slap or grab humans if they are nearby. None of us has been bitten, but having seen what Moth did to Weasley just a few minutes earlier, I don't trust him. To my amazement, Jester saves the day by charging at Moth, screaming and doing vocal threats at him. Moth and Tranquilo seem completely taken aback by this display of bravado by a female and flee far away, with Jester and several other females in pursuit. It is not often that the monkeys do favors for us (aside from providing us with interesting data, of course), and I pause to savor the moment and chuckle at the triumph of this little subordinate female over one of the burliest males around.

In the course of the next hour, there are attacks, counterattacks, and dances every few minutes. The females are far more valiant than the males in repelling the two intruders, even biting them at times, and the intruders seem far more frightened of the females in general than they are of the males. The plot thickens when Heinrich, a young adult male from Pelon group, appears on the scene. I look around anxiously to see if he is accompanied by other Pelon males. This is a crazy day! I cannot believe that after years of waiting for males I know to migrate into groups I also know, I now have observed four known males from three different social groups migrating to Abby's group on the same day. It seems like too much of a good thing. I realize that today is not going to be a great day for me to focus on training, so I radio for Nick Parker to come and switch places with the new assistant.

Everywhere I look there are monkeys dancing, threatening, forming coalitions, and double-crossing one another. The tension among the males is electrifying, as they work to gain information about their social standing in this rapidly shifting political landscape and try to enact a viable strategy. Heinrich, Moth, and Hongo engage in a three-way wheeze dance; Hongo and Heinrich maintain eye contact and circle each other nervously, as if each is trying to mount the other. Other males are attracted to the commotion: I can see Tranquilo (nervously chewing a stick), Took (angrily threatening Heinrich from a distance), and Jackson and Bentley (both directing intense vocal threats toward Heinrich). Took and Heinrich dance and sex squeak, but then Hongo and Took threaten Heinrich. Quickly the tables turn, and now Jackson and Hongo threaten the already wounded Took, who whimpers and clings to the end of a branch. Moth screams at Took, and then Heinrich attempts to dance with Moth. Jackson walks up to Took and gingerly pokes him, the way monkeys poke scary insects such as biting ants and stinging caterpillars. Then Jackson and Hongo form an overlord against Took. Jackson leans forward, grabs Took's head, and paws at him in a strange half-playful, half-aggressive way, and Took does not protest. Jackson soon grows tired of this and starts to depart.

Later in the morning, Moth and Heinrich dance together, alternating between threatening and sex squeaking. They seem highly ambivalent, and Moth's signals are made all the more difficult to interpret by the fact that he keeps stuffing caterpillars into his mouth during the interaction. Jordan, Took, and Bentley arrive to assist Heinrich in threatening Moth, but Moth calmly sits and gobbles caterpillars. No one besides Moth has eaten anything all day, presumably due to the stress of trying to figure out where they fit into this rapidly changing web of alliances. Moth's psychological poise and ability to concentrate on getting a good meal is remarkable, given the fact that coalitions of two to four relatively unknown males are continually threatening him or attempting to dance with him. Moth shows no sign of fear (now that it is the males, and not the females, who are harassing him), and he occasionally pauses to coolly shake a branch at his opponents in between caterpillar snacks, as if shooing away a pesky juvenile. When he does this, the others scatter in all directions, but each time they eventually regroup to threaten at and dance with Moth again.

At 11:00 A.M. Nick finally turns up. We agree that I will follow Moth and Nick will follow Took initially, switching to following Heinrich later in the day, so we can simultaneously collect information on the strategies of multiple key players in this drama. Nick rushes off in search of Took, whom he finds grooming with his old friend Bentley. Several females harass the grooming pair, but the males ignore them. Bentley lies on top of Took and sucks the fur on top of his head for a while, then switches to sucking his tail tip. They continue to groom and suck each other's tails for the next hour, despite repeated protests from other group members. After separating from Bentley, Took spends the afternoon becoming reacquainted with the juvenile males in the group, who squirm, gargle, and trill excitedly as they gather around him. He also forms coalitions with Hongo against Nick. Took has an excited reunion with his seven-year-old natal male friend Nut, followed by play. Although intervention from females causes Took to turn on Nut at one point and threaten him, he later has a long, relaxed grooming session with him.

Moth spends most of the next couple of hours devouring everything in sight, staying within view of Tranquilo. At one point, however, he starts sniffing leaves and then licks urine off a leaf. As soon as he licks the urine, he directs sex grunts and duck faces at Maní, who responds with soft, squeaky grunts. Maní is quite noticeably pregnant. As Jackson approaches the pair, Maní threatens Moth, who retreats toward Tranquilo. Jackson chases Moth and then attempts to strike up a dance with him.

Nick switches to following Heinrich after finishing his follow of Took. Heinrich spends most of his time avoiding or threatening males who menace him. Nut comes into view and starts to wheeze and dance. Heinrich perks up and dances with him, while grunting rhythmically, but then Hongo approaches and Nut turns traitor, ending his dance to vocally threaten Heinrich, and both Nut and Hongo chase Heinrich. Nut persistently threatens him, and then Jordan joins in the chase as well. Nut rushes up to Jordan and flings his arm around Jordan's shoulders, threatening Heinrich all the while. Heinrich puts some more distance between himself and his harassers and sits down facing them, anxiously kneading his tail tip between his two hands. Heinrich peeps three times, as if inviting grooming, but there is no one nearby to groom him. After thirteen minutes have passed and the other males have vanished, Nut and Heinrich once more break into a wheeze dance, both grunting, wheezing, and pacing. This time they manage to keep things civil for a full minute before ending the dance peaceably.

It must be emotionally exhausting for new arrivals like Heinrich to form friendships, because they do not yet understand the web of social relationships in the group they are trying to become a part of. Nut is obviously interested in getting to know him better, but Heinrich has no idea who Nut's other friends and enemies are, or who is dominant over whom in this group. At this point, he has no way of predicting what sets of social circumstances will induce Nut to support him or attack him.

As Nick and I trudge back to the car that evening, we attempt to analyze the situation. Took seems to be rapidly reintegrating into Abby's group, and he and Bentley are extremely chummy. Although Took has admirable social skills, he does not seem ambitious enough to become alpha male. Moth has phenomenal psychological poise—with the males, if not with the females—and he certainly is a good fighter. But does he have enough political sense to maintain a position as alpha? He seems to care more about eating than politicking, and we are not sure he has enough sense to get the females on his side. Besides, Moth and Tranquilo are not cooperating as much as we would expect, considering how long they have known each other. Moth would surely need a strong ally to successfully execute a takeover. Nick thinks Heinrich has the potential to be an alpha male.

During the next few days, chaos reigned, with the various males fighting, wounding one another, and switching coalition partners often. Bentley and Took were a tight but fairly nonaggressive coalition and spent inordinate amounts of time cuddling and sucking on each other's tails and toes. It seemed that Took's wounds were still bothering him quite a bit. Weasley returned in just two days but kept a low profile. Tranquilo went back to Rambo's group, and Moth followed suit the subsequent day. A few days later, both Heinrich and Nut disappeared as well. With all of the newcomers out of the way, Jackson emerged as alpha.

We had to drop Abby's group for a week during the research team's monthly five-day vacation, much to my regret, and when we returned, several males were missing and Moth was clearly alpha male. Took, Jackson, and Jordan departed shortly after Moth assumed power. Moth was not a particularly bossy alpha male: as we had expected, he took full advantage of his rank to eat as much good food as possible, and he seemed rather relaxed around the adult females. On June 11, both Diablita and Maní gave birth to new infants (Frodo and Macadamia), and on June 12 Vandal gave birth to Vodka. Moth showed no signs of aggression toward them and even groomed with the mothers in close proximity to the new infants. His rates of grooming with adult females were higher than those of previous alpha males. We began to relax and cheer Moth on as new alpha of Abby's group. Maybe this was what the group needed: a male who was too big and tough to be challenged by outsiders and too indifferent to social life to interfere excessively in other monkeys' affairs.

June 14, 2003    I am following Abby's group. Nando Campos calls on the radio to report that Rambo's group is approaching. This will be the first intergroup encounter with Rambo's group since Moth migrated from Rambo's to Abby's group. We are excited to see what Moth will do when he has to choose between allegiance to his new group and his natal group. Moth has known Pablo as alpha male of Rambo's group all his life, and he has been Tranquilo's best friend for many years. We decide that Nando will follow Tranquilo while I follow Moth.

Moth (left), the newly arrived infanticidal alpha male of Abby's group, sits by Maní, just hours after she has given birth to her infant Macadamia. The infant is killed six days later. Photo: Susan Perry.

I hear an alarm call of the type that monkeys make in response to enemy monkeys, but Moth ignores it; he is too busy eating. Hongo seems agitated and is threatening in the uphill direction and headflagging toward Moth. When Moth finishes eating, he too threatens up the hill and stands bipedally next to Hongo to get a better view. Tranquilo comes into view, and Moth directs intense vocal threats toward him. Tranquilo looks stunned by this treachery from his former best friend and curls up, kneading his tail anxiously between his hands. Moth wheezes and makes a wrinkly face at him; Tranquilo doesn't respond. Then Moth puffs up, his hair standing completely on end as he glowers at Tranquilo, and resumes threatening him vocally. Hongo joins in these threats. Tranquilo does not budge or respond directly. Instead, he nervously picks at pieces of bark and kneads his own tail, never taking his eyes off Moth.

Tranquilo turns to threaten the nearest field assistant. This does not have the usual effect of diffusing the tension between the monkeys. Moth's penis becomes erect, and he and Hongo continue to vocally threaten Tranquilo. Tranquilo retreats a bit. He is twitching all over; clearly he is highly unnerved by this encounter with Moth. His eyes dart back and forth between Moth and Hongo, but he focuses mainly on Moth. Hongo chases Tranquilo, and Moth joins in the chase, which is a long one. Tranquilo finally escapes and returns to Rambo's group when Moth stops to take a bath. Moth remains agitated for the next ten minutes and then finds a *Prockia* bush and settles in for a long meal, food calling vigorously all the while. This interaction between Moth and Tranquilo puzzles me. Why wouldn't Moth welcome Tranquilo's help? He has been a faithful ally in the past, and with so many males vying for the alpha position in Abby's group, it seems that Moth should welcome some reliable assistance.

On June 16, Jackson and Jordan return to the group, but Moth retains his position as alpha. The situation seems to be stabilizing, and I reluctantly switch to the Pelon-group team for the next day so I can work with Nando to train another new field assistant, Chris Schmitt.

**June 17, 2003**     In the morning I get a desperate radio call from Hannah Gilkenson, who is following Abby's group and also training a new assistant, Kevin Ratliff: "Macadamia is dead; I'm pretty sure it's an infanticide."

"Oh no," I groan. "Any clues as to who did it?"

"Diablita and Vandal seem really mad at Moth. They were avoiding him this morning and forming coalitions against him. Celeste has been threatening him too. So basically all of the new moms are mad at Moth. Maní is still carrying the body everywhere. She's dropped it a few times and then rushed down to retrieve it. It looked really fresh when we found her around 7:00 A.M. Maní did a single alarm call at Jackson earlier, but she alarm called at Moth for ages and ages—about twenty-five times. She threatened him and fled from him too."

"Hmm, that sounds pretty incriminating, when you consider how chummy Maní and Moth have been this week. And I've only seen a female alarm call at a group male two other times ever, one of them being right after an infanticide like this."

"Yes, Moth's my prime suspect. Can you come? I've been trying to get through but I couldn't get radio contact. I want to follow Maní till she abandons the corpse, so we can get a tissue sample, but someone should be figuring out what is going on with the males."

"Of course—I'll be right over. Where are you?"

"We're in The Anus."

"Oh, great. It just figures that they'd do their most crucial and exciting stuff in a rotten place like that." I borrow Nando's machete, apologize profusely to the bewildered Chris, and leave him with Nando to learn monkey IDs while I head for The Anus, which is a place named by a consensus of field assistants who were highly disgruntled about having to spend any time there at all. It is a seemingly endless stretch of steep, trail-less terrain covered in dense, thorny undergrowth that reaches up to three meters high. Because the visibility is about ten centimeters in front of you and it is impossible to move in most directions, it is very easy to get lost there for ages.

About half an hour later, Hannah calls again: "She's still got Macadamia. She's starting to handle the body a little more roughly, but she's not ready to give it up yet. She has been threatening and alarm calling at Moth some more, and headflagging to Diablita. She sometimes sets the body down while she eats, but she carries it everywhere she goes, and sometimes she grooms it. Oh—hang on, big fight. I'll be back in a second."

I walk faster and anxiously wait for Hannah's report. She gets back on the radio just long enough to say, "This is crazy—let's just wait till you get here for me to fill you in. Bye."

Finally I make it to The Anus, pull out my compass, and attempt to follow a path into the center of the area. Every five minutes, I try whooping. Eventually I hear answering whoops and I veer toward the sound, continuing until I find Hannah, Kevin, and Alex Fuentes. "What's happening? What did I miss?" I ask. "How are you doing, Kevin? It's kind of a rough first day on the job isn't it? Infanticide in The Anus. Honestly, it doesn't get much worse than this," I apologize.

Kevin looks exhausted, but he is thrilled to be seeing a textbook case of infanticide, one of the most rarely documented phenomena in primatology, on his first day in the field. Alex is looking somewhat the worse for wear as well, covered in mud, burs, and caterpillar-gut stains, but still energetic and ready for action. Hannah, true to form, looks as if a Hollywood director has just airlifted her from the dressing room into a jungle set: not a hair is out of place, and she looks fresh, bright, and alert, despite the fact that she has been crashing through brush all morning (at one point even dangling upside down by her feet in a vine tangle) and collecting fabulous data. It is just one of her many talents—one that has been mystifying the rest of us for years.

Hannah, Alex, and Kevin begin to fill me in. Hannah starts off, talking at lightning speed as usual, her eyes darting about to make sure she isn't missing any more monkey drama while she gets me up to speed: "When I told you I had to go, I was hearing these awful undulating screams pretty far away, and Maní and Cookie started doing alarms and threats. They ran over there, and we saw Moth and Hongo having a huge fight. Moth was chasing Hongo, and he had a cut on the mouth. It was pretty bad—the lip was slit right down the center to the chin, with the two sides flapping open, and it was bleeding a lot. Then I saw that Vandal didn't have her baby. Diablita had hers, but she was keeping it dorsal. Diablita and Celeste were threatening after Moth and Hongo. Vodka was all alone, hanging by the neck from a V in a branch, squeaking and crying. Vandal was threatening toward Moth and Hongo."

"That's bizarre! Females never let their babies go off them this early. Someone must have pulled Vodka off of Vandal," I say.

"Yeah, that's what I was thinking too. Anyway, I left Alex with Vodka while I went after Maní."

"*Escuche*," says Alex ("Listen!"), and he plays back for me the vocalizations Vodka was making. It sounds like the infant must have been in a lot of pain.

"There was saliva all over his back, but we can't tell if there are wounds," says Hannah. "Vodka fell out of the tree to the ground, and then Maní and Celeste came over to check him out, and Yasuni and Cassie came to see too. Vandal was really slow to get there, but when she did, she smelled him and hugged him to her abdomen as she walked bipedally to a stump. Then she hugged him a long time. His eyes were shut, and his head was nodding back a bit, but he was starting to cling to her. Mainly he seemed stressed out. After a long time, he managed to crawl on her back and started acting fairly normal again."

Members of Abby's group pile into a hole in a guapinol tree to rub a mixture of sap and rainwater into their hair.
Photo: Hannah Gilkenson.

Abby devours the larvae from a *Polistes* (paper wasp) nest.
Photo: Susan Perry.

Squint and her daughter Broma inspect newborn infant Eldritch, who is sleeping on the shoulders of her mother, Tattle. Photo: Susan Perry.

Tattle attacks juvenile male Quizzler, who is playing with Guapo. Photo: Susan Perry.

Tattle and her eleven-year-old daughter Eldritch (on top) demonstrate the overlord posture for coalition formation shortly after Tattle becomes alpha female of Flakes group.  Photo: Susan Perry.

Pablo, long-term alpha male of Rambo's group, embraces his co-leader Mezcla (alpha female and an ally for the past decade) as she threatens someone.
Photo: Susan Perry.

Sisters Vandal (on top) and Diablita form a coalition around the time Vandal usurped Diablita as alpha female of Abby's group.  Photo: Susan Perry.

Long-term allies
Pablo and Doble
execute a closely
synchronized attack
on the males of
Vaqueros group.
Photo: Susan Perry.

Fonz, the alpha male of Pelon group, devours a bird while adult female Miffin watches longingly. Photo: Mackenzie Bergstrom.

Guapo reaches back to help position Kola (then a juvenile) as Kola mounts him sexually. Note that Guapo is making a duck face. Photo: Susan Perry.

Abby's twins
eagerly observe her
foraging efforts.
Photo: Susan Perry.

Pablo (behind)
forms a coalition
with his son
Tranquilo, shortly
before Tranquilo
becomes alpha male
of Abby's group.
Photo: Fernando
Campos.

Fonz (on top) "overlords" with his son Power.  Photo: Hannah Gilkenson.

Two juveniles sleep in contact on a rock.  Photo: Hannah Gilkenson.

Members of Rambo's group engage in a wild bout of play on the ground.
Photo: Irene Godoy.

Two juveniles swat at
each other in play,
while hanging by
their tails.
Photo: Hannah
Gilkenson.

"*Y despues, Jackson atacó a Moth! Fue la primera vez que he visto monos peleando en serio,*" says Alex, a gleam in his eye. ("And then, Jackson attacked Moth! It was the first time I've seen monkeys fight seriously.") He reenacts the scene, having one hand represent Moth and the other, Jackson pouncing on Moth and grappling with him.

"Yes," says Hannah, "There was some big chaos involving Weasley, Jordon, Jackson, Moth, and another male. I couldn't get it all down, because I was trying to keep guard over Macadamia's body—Maní had dropped it and wandered off. But Jackson was screaming at Moth and another male, and he was so agitated he was drooling everywhere. He jumped into the tree Moth was in, landed right on top of him, and they both fell out of the tree. Everyone ran off chasing them and doing intense vocal threats. Then Vandal did a long dance with Jordan. Vodka was sleeping on her back during that."

"OK, so back to Macadamia," I say. "Is Maní still carrying the body?"

"No, she finally abandoned it about forty minutes ago. We waited a long time, to make sure she wasn't going to come back for it. When the whole group moved on, I collected it. There was a canine puncture wound in the right ribcage, and that was bleeding. The body was still slightly warm and flexible. So now that you're here, I guess I should take this home and get some tissue in preservative, right? I'll bury the body in case some museum wants to use the bones for research or an exhibit."

Jackson saunters up to us. He has cuts on the right wrist and elbow, on the left shoulder, and on the forehead. His face is spattered with blood. Five monkeys are threatening him, but he seems unconcerned. Hannah snaps a few pictures of him and then packs up her bag to head back to the car and to town.

I make my way through Abby's group, inspecting everyone. There are several new wounds. Diablita has a cut on her left shoulder that is bleeding. Weasley is favoring his left hand. I can't find Moth for the rest of the day. It appears that Jackson has been successful in driving him out and retaking the role of alpha male.

As it turns out, it is possible for Kevin and Chris's rough first day in the field to get even worse. When we all squeeze into the car again at the end of the day, we are exhausted and dying for dinner. The driver begins trying to turn the car around, a difficult feat on an unlighted road full of muddy, rocky gulches that are so filled with water that you can't tell how deep they really are.

"Keep backing up!" yells someone from the backseat who is looking out the rear window.

"Are you *sure?*" asks the driver skeptically.

"Yeah, go for it," retorts the confident backseat driver.

There is a lurch and a crash, followed by the futile spinning of wheels as the driver strains to get the car out of the ditch, and then an appalling explosive sound just before the engine dies completely. Attempts to re-start the engine fail. We all pile out of the car in alarm and pull out our flashlights to see what is going on. The exhaust pipe has been driven deep into a muddy bank. When we manage to dig it out a bit, we are aghast to discover that it has been bent at a 90-degree angle and is packed full of mud and rock. It looks as if an archaeologist has been us-ing it to try to collect pollen core samples. While Alex and Nando try to dig the wheels out of the mud with their machetes, others of us collect rocks to pile under the wheels for traction, and I break various compo-nents of my Swiss Army knife trying to extract tightly compacted debris from the exhaust pipe. We go at this for thirty minutes or so before Alex shoos us, the incompetent and interfering gringo academics, aside so he can work in peace.

Despite the fact that we have all owned and driven cars, it is nonethe-less undeniably true that Alex and the other Costa Ricans on the proj-ect are far more capable than we are when it comes to fixing cars—or fixing just about anything else, for that matter. We are all covered from head to toe in slime and filth from crawling under the car. We turn off our flashlights and lie down in the middle of the muddy road to rest, star-gaze, dream of our eventual dinner, and discuss the day's events. Even-tually the conversation turns to theory (something we actually know about, unlike auto repair), and we try to predict the fates of the infants.

"So what do you think? Does Vodka have a fighting chance?" asks a voice out of the dark. "Is infanticide common in capuchins?"

"We talked about infanticide in my primatology class!" responds an-other voice. "It's really common in primates, and it's a male reproduc-tive tactic. Males who are migrating into new groups kill the infants so that the females will come into estrus sooner, and then they mate with them. Right?"

"That's right," I say. "That's Sarah Hrdy's hypothesis—the sexual selection hypothesis about infanticide.[2] She thought it up while watching infanticides in Hanuman langurs. Infanticide is a really widespread occurrence in mammals generally, not just in primates, but you wouldn't expect to see it in all mammals, just in species and populations with the right demographic conditions. It's quite common in lions, for example, which have a similar social structure to that of capuchins. Lions are female philopatric, and pairs or trios of males who are often related migrate from group to group, throwing out the males, killing the infants, and mating with their mothers."

"What are the right demographic conditions for capuchins? Do we have them here?"

"Well," I say, "infanticide is expected to be most advantageous in species in which the tenure of the alpha male is short relative to the amount of time required from conception to weaning. So basically, the shorter the male tenures are, the more infanticide you are likely to see. If, for example, there is typically a takeover in leadership every one and a half years, the normal interbirth interval is two years, and the new male has effected the takeover at a time when the females have brand new infants, then a male can hardly afford to wait until the newborn infants are mature for females to come into an ovulatory state. If he does so, he may forfeit his only chance to reproduce. His best bet for leaving a genetic contribution is to kill the infants and mate with the mothers. In the past, we have had long alpha tenures, and we didn't really see any infanticide here except for one case in which we suspect that the mother didn't mate with the alpha. The other two cases we have seen have been new alphas killing infants that presumably were sired by the previous alpha. Right now, with all the migrations and changes in alpha male status, I'd say these infants are at very high risk."

"I don't get it, why does killing the infants make the moms ready to mate sooner?" asks a voice.

"The thing is, nursing inhibits ovulation, the release of the egg into the ovarian tubes where it can be fertilized. Nipple stimulation triggers the release of oxytocin, and this is what inhibits ovulation, through a chain of hormonal interactions. The more frequently nursing occurs, the more inhibiting its effect on ovulation. Infants nurse at higher frequencies when they are younger, so the younger the infant is, the more its presence will delay its mother's return to a fertile state. A female can be nursing at very low rates and still conceive, though. So it only pays a male to kill quite young infants."

"Still, it seems like it would only help the killer if the female mated with him rather than someone else, and why would she want to mate with some jerk who just killed her baby?" says an outraged female voice in the dark.

"Exactly. Females are not just passive players in the game of male reproductive strategies. They have evolved many counterstrategies of their own to cope with this problem. Females often ally themselves with strong males who can help defend them and their infants from intruders."

"Maybe that's why all the females are sucking up to Fonz all the time in Pelon group. There's *no* one tougher than Fonz," says one of Fonz's loyal male fans. Fonz is indeed a hulking brute of a male and has held the alpha rank for many years. I can sense certain women rolling their eyes, even in the dark.

"Yes, but the females don't always depend just on males," I hasten to say. "Females may band together with other females to resist immigration by new males, or to defend their vulnerable infants from a male who has successfully taken over. Just like Diablita and Vandal did today, and like Abby and Squint did in the old days. In many species, females who are already pregnant with another male's infant will come into postpartum estrus and mate with the new male, as a way of confusing paternity. Depending on the exact nature of the psychological mechanism involved in triggering infanticide, this tactic may fool the male into thinking that the offspring is his own, so that he will not kill it.

"But—to get back to your question of why the female would mate with this jerk—once the infanticidal deed has been performed, there is typically no advantage to a female in refusing to mate with the killer. Unless she has a better option, she would be sacrificing her own reproductive interests to punish the male, and such tactics would be selected against. Infanticide is most common in species in which single males are able to monopolize reproduction, either by intensive mate guarding or by being the only male in the group. Mating with a subordinate male or a male from another group, who would be ineffective in defending the infant from further infanticidal attacks, the female would be jeopardizing the chances of survival of her future offspring. Chances are, the male who just killed her previous offspring is also the male who is best qualified to protect her future offspring from attack. So if you assume that the males can, at least with a better-than-chance level of certainty, determine which offspring are theirs, then mating with the killer is probably her best bet, unpalatable as it sounds."

"Eeeewww, but how can they be sexually attracted to someone who has just done something so awful?" someone asks.

"I don't know," I laugh. "Figuring out what it actually feels like when monkeys make these decisions is too ambitious a research question for me—I don't think it's possible to test that! I'm just going to stick with testing straightforward predictions from behavioral ecology, assuming these females have been evolved to do whatever is in the best interest of the survival of their future offspring."

"So what's going to happen to the rest of the babies?" asks another field assistant.

"I can't predict the future," I say. "If I knew who was going to be alpha and for how long, I'd be more willing to hazard a guess. I'm guessing Jackson will turn out to be the father of most of this year's birth cohort, though probably not all of them. If he stays alpha for the next year or so, I'd guess they have a decent shot at survival. He really did a number on Moth today. But Jackson hasn't exactly done a good job of staying put and keeping his rank over the past year. I'm worried. We'll see."

*"Listo! Vamanos!"* says Alex. ("We're ready to go!") We all cheer and turn on our flashlights to gather our things and pile back into the car. The car will definitely need a trip to the mechanic, but Alex has jerry-rigged it well enough to get us home to dinner, which is all we care about at the moment. It has been a long day.

Our high hopes for Jackson's political career and peace in Abby's group don't pan out. Four days later Moth returns, and we see that he too has several new wounds, most likely from the fight with Jackson. He does not seem to be making a play for the alpha position anymore but stays on the edge of the group. Weasley, Jordan, and Jackson cooperate in keeping Moth at bay. Took and Bentley are still spending much time cuddling and sucking on one another's body parts. It is no longer clear which male is alpha.

The plot thickens further when, on June 24, Punto, Heinrich, Power, and Alamasy arrive from Pelon group to stalk Abby's group. This same day, Eldritch gives birth to her baby Eggnog. Still no one assumes the alpha position. Moth joins the Abby's group males in trying to repel the Pelon intruders, but they are persistent. Many males (Moth, Punto, Power, and Heinrich) are wounded during this power struggle. Jordan leaves the group when the Pelon males arrive and never returns, aside from a couple of brief visits later in the year. Jackson's attendance in Abby's group is sporadic after Jordan leaves, and Jackson eventually departs for good as well. Punto inspires more fear than any other male, though he does not seem to make an effort to be central in the group, or to win the females' favor. Punto shows a complete lack of fear himself, even when confronted by coalitions of five adult males. Diablita and Vandal, the two top-ranking females, are particularly vehement in their attempts to repel Punto from the group, but they are unsuccessful.

At the end of June I have to return to my work in Germany, at the Max Planck Institute for Evolutionary Anthropology. Every day I anxiously check e-mail for the latest news about the monkeys. I learn that when the team members returned from their five-day vacation at the end of the month, Cricket (Celeste's infant) was missing and Punto seemed to be alpha male. Tranquilo (Moth's sidekick from Rambo's group) paid a visit to Abby's group and was chased away by Moth, Hongo, and Jackson. Another Pelon male, Gandalf, had joined Abby's group and was just as unpopular with the group's females as the rest of the Pelon males were. Over the course of the next month, Hongo was the only former Abby's-group male who was in regular attendance. Weasley and Bentley were present about half of all days, and Jackson and Took were around less than a quarter of the time. In the latter half of the month, two more babies were born: Opie gave birth to Ox, and Jester gave birth to Jinx. Increasingly, the group had trouble maintaining cohesion and started to split into subgroups.

In the last week of July, while the team was again away from the field, Jester's infant Jinx disappeared. It was so frustrating to have missed yet another probable infanticide—and depressing to lose another focal subject. During the month of August, Punto was still clearly alpha male. But unlike all other alpha males I had known, who reveled in the attention received from females and took full advantage of the alpha's groom-

ing privileges, Punto seemed uninterested in such activities. He continued to be absolutely fearless and impulsively violent, to the extent that
he seemed to thrive on aggression. Eldritch, one of the new mothers, was
particularly afraid of him; she would avoid him and would suck her tail
when he came near. Field assistants Laura and Chris were thus not particularly surprised when they arrived on the morning of August 23 and
found Eldritch's infant Eggnog had been severely wounded. There was a
large cut on his back, deep enough to expose his internal organs, and his
legs were paralyzed, so that he could not cling with his feet and needed
his mother to help him hold on while she carried him. Eldritch spent all
morning tending to her infant, nervously monitoring Punto's activities,
avoiding him, and alarm calling at Punto continuously. It seemed pretty
clear whom Eldritch was holding responsible. By 10:30 A.M. the baby
was starting to turn bluish and have trouble breathing, and by 11:00
A.M. Eggnog had died. Eldritch and Jester left Abby's group the next
day, and by the end of the week Eldritch; her mother, Tattle; Jester; and
their descendants had moved to the banks of Río Pijije and formed their
own group, separate from the rest of Abby's group. Initially Jackson and
Jordan accompanied them, but eventually they were replaced by Heinrich and Alamasy (two of the Pelon males), and Heinrich became their
alpha male.

Weeks later, on September 10, after a two-day gap in observations,
Opie's baby, Ox, was found with a deep gash on his back. Punto, Heinrich, Alamasy, and Gandalf were the only adult males present in the
group on that day, but of course it is possible that a different male had
inflicted the wound on the previous day. The next day, Opie was discovered with a bite on her tail and an injured foot. Had she been wounded
while defending her infant from further attack? At 10:14 that morning,
Hannah heard terrible screams and ran over to see Vandal and Punto
hanging from their tails, screaming and grappling with each other. Monkeys on the branch above them were also screaming. Punto bit Vandal's
infant Vodka, and Vodka was thrown from the tree and landed on the
ground by her sister Cookie. Vandal and Punto then fell to the ground as
well, and Vandal, Diablita (who was carrying her own infant, Frodo, on
her back), and another female chased Punto away.

Vodka had a bloody face and a cut left ankle that was bleeding profusely. The hurt infant continued to scream. Cookie seemed highly agi-

tated and shook branches at Hannah and alarm called at her. Vandal did not run to comfort Vodka immediately but kept threatening in the distance toward Punto. He slowly returned, and began to sex squeak and grunt toward Vandal, who avoided him. Hannah could now see that Punto had a new wound halfway up his tail. He stopped to lick the blood from it and groom it, then grunted at Vandal again. Vandal was not the slightest bit interested in dancing with Punto, and she repeatedly alarm called at him. Vandal collected Vodka, who was able to ride dorsally but could not grasp with the recently wounded foot. Vandal continued to stare at Punto and alarm call at him repeatedly for the rest of the day, just as Maní had done toward Moth when Macadamia was killed and Eldritch had toward Punto when Eggnog was killed. Cookie joined Vandal in alarm calling and fleeing from Punto.

The following day, September 12, proved to be one of the most pivotal and perplexing days in the history of Abby's group. Hannah and Chris found Hongo, Punto, Power, and Gandalf at one of their usual sleeping sites, and saw that Gandalf had severe new wounds. Chris followed the four males until 8:00 A.M. They seemed agitated, as if patrolling the area, and although Hongo rejoined the main group later in the day, the Pelon males never came back to Abby's group. When Hannah found the group's females, they were with Tranquilo and two subadult males from Rambo's group—Elmo and Duende. Diablita and Vandal seemed to have no objection to the presence of these males, which was quite a contrast with their attitude toward the Pelon males. Diablita and Vandal had been vehemently united against the Pelon males (and against Punto in particular) for the past several weeks, often putting themselves in great danger to defend each other's babies from possible attack. The infants Vodka and Ox were still alive, though neither was doing particularly well. Opie had to carry Ox ventrally most of the time, and Vodka's right foot was not usable. Later in the afternoon Diablita did have a huge fight with Elmo, in which she solicited aid from Tranquilo (who sided with Elmo instead) and then chased Tranquilo as well. But the Rambo's-group males were there to stay, and the Pelon males never returned.

It will always be a mystery to us how three relatively nonaggressive males—Tranquilo (who was about eleven years old), and Elmo and Duende (both seven-year-old subadults)—became the sole male residents of Abby's group without a single battle, when thirteen fully adult males, all capable fighters and (for the most part) experienced politicians, had fought among themselves for two years straight for this privilege. And why did none of those vanquished males ever return to try again? Hongo and Weasley made a few sporadic and nonaggressive visits to Abby's group and then vanished forever. Heinrich became alpha male of Flakes group (the subgroup formed by Tattle, Jester, and Eldritch). And the rest of the Pelon males, except for Punto, ultimately returned to their natal group. Jackson and Jordan have since been seen wandering the forest together on their own, as have Bentley and Took. Moth returned to Rambo's group, his natal group. At the time of this writing (September 2006, three years after Tranquilo assumed the position of alpha in Abby's group), there have been no challenges to Tranquilo or noteworthy immigration attempts (though Jackson and Punto paid one more brief visit in October 2003).

On Tranquilo's fifth day as alpha male, Ox died from the wounds he had suffered a week previously. Opie continued to carry the body for three hours and then abandoned it. The previous year's birth cohort eagerly nursed from her once Ox was gone, and she seemed more than willing to permit this, even presenting her nipple to two-year-old Nobu. Vodka was still limping from his previous wound. Relations between Tranquilo and the Abby's-group females (Diablita and Vandal, in particular) became progressively worse, and Diablita often threatened Tranquilo.

On September 20, Vandal was discovered in the morning by field assistants Laura, Alex, and Heidi Ruffler with new wounds: a broken leg, a cut arm, and a split lip. Her finger had been injured the day before. Perhaps because of the stress of these injuries, she left Vodka alone or with her daughter Cookie most of the day. Vodka regularly gargled to Tranquilo from a distance but appeared lethargic and skinny and spent quite a bit of time grooming his wounds. When Tranquilo came near Vodka, Cookie began alarm calling and collected Vodka; the rest of the group then threatened Tranquilo. Around 10:30 A.M., Cookie set Vodka down while she foraged. She occasionally alarm called. At one point she became absorbed in foraging and strayed to fifteen feet away from Vodka. Tranquilo, who was watching all of this, quickly and quietly sneaked up to Vodka and bit the infant. Cookie alarm called and ran back toward them. Two other monkeys, who could not be identified in the chaos, also ran to Vodka's assistance and attacked Tranquilo. Vodka, Tranquilo, and Vodka's three supporters fell from the tree, dropping five meters to the ground, and Tranquilo was chased to the edge of the group. Diablita threatened Tranquilo, and Tranquilo and Elmo formed an over-lord against her. Cookie alarm called continuously and ran around aimlessly; Vodka was on the ground looking dazed, but was still alive. Vodka looked at Cookie and called to her, but she did not pick him up. He had a new bite wound on his wrist, which seemed to be broken. For the next half hour Cookie alarm called hysterically and remained with Vodka but did not pick him up, despite Vodka's pitiful lost calls, coos, and peeps. Eventually Vodka tried to crawl toward the rest of the group, but with a broken arm and a severely injured leg, he kept toppling over. By noon he was beginning to acquire the bluish coloration that had become all too familiar to us as heralding the imminent death of an infant. Laura left Vodka with Alex while she went to look for Vandal. She found Vandal about fifty meters away, foraging. Vandal was moving quite slowly and appeared to be in very poor health herself. It is completely abnormal for a mother to leave such a young infant, particularly when it is in distress, but perhaps Vandal was in such a bad state that she could not cope any more. Vodka spent the last two and a half hours of his life alone, though in auditory range of the group, and died at 1:30 P.M.

Now Diablita's baby Frodo was the only surviving member of the seven-infant birth cohort. I continued to check e-mail messages daily, praying each time to see yet another "Frodo lives!" in the subject heading. Vodka's death appeared to have had a profound effect on the group dynamics. Vandal seemed to be deeply depressed and continued to lose weight. Diablita distanced herself so much from the group's males that she was difficult for the team to find. With the alpha female always on the periphery, and the dynamic duo of Diablita and Vandal no longer right in the thick of things, interfering with the males' activities, the group had quite a different character and lost its cohesiveness. Abby's group had always been notable for the strength of its female-female bonds and the feistiness of the females when confronted with male obnoxiousness. Now that Vandal was deeply depressed and Diablita was constantly on the run from Tranquilo, the Vandal-Diablita bond that had been so firm for so many years was weakening.

Every member of the research team was impressed by the valiance and determination of Diablita in defending Frodo from Tranquilo. He made some attacks on Frodo but Diablita fought him off, even wounding him. Capuchin mothers are normally quite relaxed about letting their infants run off and play, but Diablita was hypervigilant, retrieving Frodo whenever Tranquilo came near. The poor infant had hardly any playing opportunities. Diablita managed to keep Frodo alive until he was six months old. As the weeks ticked by, we began to relax our guard a bit, hoping that maybe he had passed some critical age threshold beyond which Tranquilo would consider him too old to be worth killing. We lost track of the group between December 10 and December 17, when the team needed to catch up on collecting data on other groups. Now that Abby's group had fissioned into two parts, we no longer had sufficient staff to follow every group every day. When they returned to Abby's group on the December 17, Frodo was gone.

This event was to change Diablita's life profoundly. While she had been off on her own, hiding Frodo from Tranquilo, the other females (including Vandal) had been busily forging a solid supportive relationship with their new alpha male. To our great dismay, Vandal now turned on her sister and longtime ally, soliciting Tranquilo's aid in overthrowing Diablita for the alpha-female position. Just as Vandal had seemed depressed following the death of Vodka, Diablita was very subdued after the loss of Frodo, and it was easy for Vandal to push her advantage. This drama dragged on for months and sometimes involved intense physical fights between the two females. It was heart-wrenching to watch Diablita cowering, trembling and kneading her tail, when confronted by the new Tranquilo-Vandal coalition—especially after she had put her life on the line to try to save Vodka from Tranquilo and Punto on so many occasions. There were still times when the two females would cease hostilities and groom as before, but these moments grew increasingly rare. All of the females but Diablita mated enthusiastically with the new males, and even Diablita eventually mated with them (though there was a striking coercive quality to her early sexual relations with these males that was atypical of the normal male-female sexual encounters). In the course of the next several months, all the females conceived infants. Maní's infant died of a birth abnormality and she had to try again (the next time was successful); all the other infants have survived to the present day.

Having observed how devastating male migrations can be to females and their offspring, it now made perfect sense to us that females would normally be staunch supporters of the alpha male and vehemently opposed to males attempting to immigrate. It is clearly to the females' advantage to keep an alpha male firmly in his seat, at least until father-daughter inbreeding becomes unavoidable. The most stable group configuration seems to be a strong alpha with two to three loyal subordinate supporters. A strong female-female alliance can help the alpha maintain his position against his group's subordinate males, and the male-male alliance can fend off intruder males from other groups.

In the absence of infanticide, female wounding rates and infant mortality rates are quite low. In Abby's group under stable demographic conditions—that is, when there has been no change of alpha male—infant survivorship has been 80 percent (excluding deaths known to be due to congenital defects). During periods of demographic instability, survivorship dropped to 48 percent. Linda Fedigan found even more extreme effects of male takeovers on infant mortality in her eighteen-year demographic survey of the Santa Rosa capuchins: in her study, 82 percent of infants died in the year following a takeover, whereas only 12 percent died in demographically stable years.[3]

But were infanticidal males really improving their reproductive success by killing infants? This would be the case only if (a) the killers succeeded in shortening the interbirth interval of the females, and (b) the females subsequently mated with them. In Abby's group, I looked at the interbirth intervals for the seven females whose infants were victims of infanticide, comparing their intervals between the surviving infant preceding the birth of the killed infant and between the birth of the killed infant and the subsequent infant born. On average, the interbirth intervals were sixteen months longer in the absence of infanticide, which suggests that the males really were enhancing their chances to sire an infant (Linda Fedigan found a twelve-month advantage to infanticidal males in her study). The average time to conception after an infanticide was five months in Abby's group.

Hastening the time to conception confers a reproductive advantage on the male only if he subsequently gets to mate with the female, and only if the infanticidal male was *not* the father of the infant killed. This is a trickier hypothesis to test, because it requires genetic testing of the infants, and there also must be an opportunity for the male to mate with the victim's mother. In Linda's study, she was able to confirm that infanticidal males stayed in the group long enough for the females to conceive another infant, though her research group was not able to accurately measure the reproductive success of the killers in the year after a takeover. Because the bodies of infanticide victims are often lost, and such young infants rarely defecate, we had few opportunities for genetic testing of these individuals. Also, there were some cases in which we could not be absolutely certain of the identity of the killer, though circumstantial evidence (like alarm calls by the mother toward a particular male during the aftermath of the attack) could be used as a clue.

For the three cases in our database in which the attack leading to death was witnessed, the killer was the alpha male. There have been only two cases in which we have known the identity of the killer, the father of the victim, and the father of the next conceived infant. In one of these, Tranquilo killed Jackson's infant, and Tranquilo sired the mother's next offspring. In the other case the killer was not the father of the victim nor of the next conceived offspring, for he was evicted from the group before the female came into estrus again. There were four cases in which we genetically determined the father of the infanticide victim, and in none of these was the killer the father. In the other four cases from Abby's group in which we could conduct no genetic testing, it was possible to exclude the killer as a possible father by virtue of the fact that he was not a group member at the time conception occurred.

Thus, white-faced capuchins have joined the growing list of primate species in which infanticides by males fit the patterns predicted by Sarah Hrdy's sexual selection hypothesis for infanticide.[4] Considering only the empirical evidence gathered during the thirty-two years since it was proposed, it would seem to be as well established as a hypothesis can be in the behavioral sciences. Yet the controversy surrounding it refuses to die. Primatologist Volker Sommer calls the situation a "holy war" and has analyzed the ideological (i.e., nonscientific) issues involved.[5] For many scholars, especially those trained in the social sciences rather than the biological sciences, a behavioral pattern as selfish and destructive as infanticide simply cannot be an adaptation: it must be a by-product of adult male-male aggression, or a pathological response to human-caused changes in a population's habitat, or perhaps not a real pattern at all but a figment of observers' macabre imaginations.[6] Of course, these scholars see the supporters of Hrdy's hypothesis as being the ones who are ideology driven.

We got a bitter taste of the issues involved when we submitted, with

Julie Gros-Louis, a short article describing the first three apparent infanticides at Lomas, for publication in a primatological journal. Admittedly only one of the cases (Jackson's killing of Einstein) was as clear-cut as most of the others described in this chapter. However, we included all the hedges and qualifications required of scientists when evidence isn't airtight. We were shocked when all five referees, using abusive and contemptuous language, urged the journal editors to reject our manuscript. One wrote that "to argue that these infant losses are due to males killing infants is a leap of theory (or faith if you prefer)." A second referee compared the widespread acceptance of Hrdy's hypothesis to Western biologists' pre–World War II acceptance of eugenics (the goal of "improving" humanity through selective breeding).

Even when an issue is controversial, scientists expect their manuscripts to be evaluated by referees from both sides of the controversy, which clearly hadn't happened in this case. We asked the advice of a senior primatologist who had extensive experience in the "infanticide wars," and he pointed out that "infanticide denial" is almost exclusively an American phenomenon and suggested we might have better luck submitting the manuscript to a European journal. He was right: our article eventually appeared in the Swiss journal *Folia Primatologica*, after we incorporated minimal (and constructive) changes suggested by reviewers.[7]

CHAPTER 9

# Kola and Jordan: Lethal Aggression and the Importance of Allies

Although capuchins spend the vast majority of their time inter-acting with the same individuals day after day, all monkeys—including the females, who form the stable core of the society—can expect to go through some periods of instability such as those described in the previous chapter, in which they are forced to form new relationships and have to cope with the cognitive challenge of remembering the reputations of many relatively unfamiliar males. Encounters between groups provide a forum in which males can gather information about one another—information that they will store away for future use, when they may need to decide whether to ally themselves with a certain male and comigrate or fight with him. Decisions about whom to migrate with or whom to admit as a group member are perhaps the most important decisions a capuchin monkey makes, and errors can prove deadly.

March 20, 1997     Todd Bishop and I are conducting a "dyad follow" of a pair of adult males. The goal of this phase of my research project is to document the nuances of male-male strategies, including long-distance communication, and Todd and I are in radio contact so we can tell each other what our respective focal animals are doing. My focal for the afternoon is Rambo, a young adult male who is of the right age to be migrating for the first time. In fact, he has been making frequent "field trips" with his best friend, Q, to visit other groups. Todd's focal is Feo, an exceedingly decrepit, freckled, scarred, and scrawny old male. Rambo is resting near Pablo, and Feo (aka the Crunchy Bug Expert) is efficiently searching through leaf litter for grasshoppers, far from the other males. Suddenly Pablo perks up his ears, sits up, and emits an intense vocal threat. Rambo looks in the direction in which Pablo is facing and threatens also. Pablo dashes toward the river and Rambo follows. "That was Pablo and Rambo. We may be starting an intergroup encounter," I radio to Todd.

"Yeah, I heard that," says Todd. "Feo looked toward the threats, but he's not going that way." Rambo threatens me and headflags toward Pablo. Pablo ignores him and keeps running toward the river. Rambo follows, and they both charge pell-mell along the ground though a mass of bromeliads.

"What's Feo doing? Is he going to fight?" I ask Todd as I chase the males.

"Heck no, he's duck facing to Delilah and Mezcla," laughs Todd. Feo is repeatedly making duck faces (part of the sexual solicitation routine) to both females and trying to get them to follow him in the opposite direction from that in which Pablo and Rambo had run. When the females keep ignoring him, he lies down resignedly in the middle of the river for a bath.

"Feo heard that. He's running to hide," says Todd in response to some screams. "Wait, here he comes again. He's just grabbed Simpson [Delilah's baby] and is running away with her on his back. A bunch of females and juveniles are following him. They are headed upstream, toward where we were."

I see Pablo and Doble (a male who looks almost identical to Pablo and is his closest male friend) standing close to each other, glaring at some screaming enemy males from the Vaqueros group (the "cowboys"). We have had frequent trouble with this group lately. The Vaqueros' fighting contingent consists of an old, large, and somewhat cowardly male, Dusty, and three virtually identical younger males (Lefty, Lucky, and Ned) whom we suspect are his sons. These three young males are holy terrors: though they are not particularly large or muscular, they seem absolutely fearless and impulsively violent. As usual, the young males are on the front line, and Dusty, whose hair is standing completely on end, is lurking at a safe distance in the background, clacking his teeth together loudly and occasionally peeking out from behind a tree trunk at me. Ned charges at Pablo and Doble and they flee, even though Ned is far smaller than either of them. The Vaqueros chase them for quite a distance, and then Pablo spins about and faces the enemy. Doble stops too, within five body lengths of Pablo, and he faces the Vaqueros with him.

Lefty is the only one near them, and he is starting to make odd sounds as Pablo and Doble calmly look up at him. Doble eases himself into the river for a quick bath while he glares up at Lefty. (Monkeys often get overheated by the exertion of intergroup encounters, and time-outs for cool baths are common.) As Lucky continues to make strange warbling noises toward Pablo and Doble, Pablo slowly advances on him, walking on the ground. Doble follows Pablo, keeping within one body length of him, truly looking like his shadow. Tranquilo, a juvenile male from Rambo's group, follows them at a distance. Pablo and Doble slowly penetrate into the midst of the Vaqueros males.

Suddenly the Vaqueros flip out, screaming and threatening, and they all resume chasing Pablo and Doble. One of the Vaqueros males nearly falls out of the tree in his excitement. Pablo and Doble are so angry that they hop up and down as if they are on pogo sticks while threatening up at their pursuers in the trees. Pablo and Doble maintain physical contact with each other as they scream at the enemy. Pablo leaps on top of Doble in the overlord position, glaring at them. Ned and Lefty move in a bit closer, but Dusty and Lucky keep their distance, apparently a bit nervous about my presence. The Vaqueros stop screaming and resume making those warbly vocalizations—a sound that I have never heard outside the context of intergroup encounters. Keeping an eye on them, both Doble and Pablo submerge themselves in the river for a bath. Tranquilo begins to dance with Ned, pacing and making duck faces. Doble follows suit, pursing his lips in a duck face and twisting his head about while maintaining eye contact with Lefty. Such interactions are common in intergroup encounters: often it seems that the males cannot decide whether they want to kill each other or make new friends, and these sudden reversals in mood and tactic are one of the most exciting aspects of the encounters for us. I report this turn of events to Todd.

"Feo's still carrying Simpson, and he's still running with the females," replies Todd, a tune of amused exasperation in his voice. He seems irked to be missing out on the action.

The face-off between the two groups continues for several more minutes, with no clear winner. Sometimes the Rambo's-group males chase the Vaqueros, and other times the tide reverses. All of the males are constantly vigilant and experiencing rapid mood swings as their evaluation of the situation changes according to the composition and behavior of their side relative to the opponents' side. Sometimes they stop fighting to bathe or to strike up dances with the enemy. Todd reports that Feo is asleep near the females, while Pablo, Doble, and Rambo continue to fight.

When Q (the youngest adult male in Rambo's group) appears for the first time in the fight, this seems to turn the tables in favor of Rambo's group again. The Rambo's-group males chase the Vaqueros several hundred meters downstream. I run as fast as I can, but I cannot quite keep up. I hear screams ahead of me. There are now two fights going on in different places, as not all of the Vaqueros have retreated to the same place. Matters become very confusing at this point, as the action is too scattered for me to follow it. Screams come from multiple directions.

When the fighting is over, I find Pablo and Doble lurking where most of the action occurred, intently sniffing branches. Q comes over to them and sniffs the same branches they sniffed. Branch sniffing is extremely common right before and after intergroup encounters; presumably the monkeys gain some useful information about who is in the vicinity. Perhaps the urine also conveys some information about other monkeys' competitive status, but we do not know for sure. The three males remain at the site of the encounter for a few more minutes, vigilantly looking downstream, and then they return to the females, juveniles, and Feo. The entire encounter has taken approximately one hour.

What was this all about? In the early years of primatology, it was assumed that battles between groups were geared primarily toward defense of food resources. In all primates, with the possible exception of callitrichids (marmosets and tamarins), females are primarily responsible for parental care, and females always bear the burden of pregnancy and lactation, which are energetically expensive and completely unavoidable costs of reproduction. Males, in contrast, may invest as little as a single copulation in the production of an offspring. This sex difference implies that female reproduction is limited by access to food, whereas male reproduction is limited by access to mates.[1,2] Primatologist Richard Wrangham deduced from this principle several hypotheses about primate social organization (see Chapter 6). Whenever food sources occur in large, monopolizable patches, females are expected to form linear, kin-based dominance hierarchies within social groups. They are also predicted to take the primary role in defense of food resources against other groups. If males participate, they will do so in the role of "hired guns," performing a service to females in exchange for benefits such as matings.[3]

Chapter 6 describes how white-faced capuchins conform to this model in some ways but deviate from it in others. Here I will focus on the most clear-cut and striking deviation: capuchin females' failure to participate in intergroup battles, except on rare occasions that usually involve confrontations between two recent fission products. When females detect the presence of another group of capuchins, they silently grab their infants and run in the opposite direction, leaving the males to do the fighting.[4] Because females retrieve their infants in only one other context— the sighting of an extremely large raptor—it seems that the females are fearful that their infants would come to harm at the hands of foreign males.

If males are fighting to defend their group's access to food resources, then it seems that the winning group would stay at the site of the conflict and continue to forage. But that is not what happens. Typically when two groups meet, the females of both groups flee quite some distance in opposite directions, while the males stay and fight. One group of males eventually displaces the other, but then no one stays for long or feeds at the site of the conflict. Both sets of males wander back to their females after the conflict has ended. Of course, it could be that the males are fighting for access to this land (and its food) in the more distant future, rather than for immediate access. However, capuchins are not strictly territorial: their home ranges are far too large to be effectively defended, and 50 percent to 80 percent of most groups' ranges are "overlap zones" with one or more groups. So this hypothesis seems unlikely as well. What seems most likely is that males are defending their access to females. By fighting against the other group's males, they are not only keeping them from gaining information about, or immediate access to, their females but also showing how tough and hostile they are, lest the males be contemplating migration. Usually the males on one side put up a united front against the enemy males.

Immigration of males often results in chaos and some bloodshed. Occasionally very young males can immigrate without causing a major disruption in a group, particularly if they migrate alone: for example, I have witnessed ten single migrations, and in only one case did the immigrant execute a takeover; in the rest, there was minimal disruption of the group's social dynamics. But all too often—in well over two-thirds of the male migration attempts observed at Lomas—males elect to migrate with one or more male companions. These group migrations are more disruptive, and the new immigrants sometimes kill group members and overthrow the alpha male. Similar patterns have been recorded by Kathy Jack and Linda Fedigan at Santa Rosa, where males have migrated in the company of other males in 71 percent of emigrations, and four of the seven hostile takeovers witnessed were accomplished by coalitions of males rather than by single migrants.[5,6,7] It is not only adult males that receive wounds or are killed; often females and their infants are affected as well, as we just saw in Chapter 8. When resident males prevent foreign males from migrating, they are not only protecting their future access to mates but also fighting to preserve their own lives and the lives of their existing offspring.

Of course, not all males have an equal stake in defending their group. Alpha males have preferential access to mates, and they also have the most offspring in the group. Therefore it might be expected that they would be more active in defending the group than other males, particularly if the other males are nonrelatives. Indeed this is the case. Alpha males participate in about 90 percent of all intergroup encounters. Dusty, of Vaqueros group, is a notable exception, though it is possible that he participates more often when not intimidated by the presence of unfamiliar human observers. In all five of our habituated groups, alpha males take the lead. Most subordinate adult males participate in 60 percent to 85 percent of encounters (Feo set a record low of 25-percent participation, much of which was merely vocal participation from a distance).

How do males decide whether or not to participate in a given encounter? Looking back, it made sense that Feo was so lazy at that stage in his life. He had no surviving offspring in the group (as we learned later from genetic analysis), nor did he have much of a shot at mating success in the future, given that he was quite old and decrepit and definitely no match for the other males in the group. Migrating to a new group would have been extremely dangerous for him, because he was in no condition to fight other males. Probably his best shot at sneaking a mating was when the rest of the males were busy battling the enemy. But whereas Feo was fairly predictable in his cowardice, other males are less so, and I am often puzzled to see a normally brave male skitter silently off with the females when a fight breaks out.

Do males punish one another for lack of participation? This question is a hard one to answer. I have never seen any flagrant punishment of monkeys, such as Feo, who fail to be helpful in intergroup battles. But of course if the punishment consists of lack of tolerance for matings occurring much later, it would be hard for us to link the punishment with the "crime."

It seems to me that males are more likely to help one another in intergroup encounters if there are other males observing their response when an enemy group is first detected. However, it is nearly impossible to obtain good data in such situations because the monkeys typically realize that an intergroup encounter is imminent before I do. I designed a playback experiment to test the effects of audience presence on males' response to playbacks of recordings of intense vocal threats by foreign males. This test would have compared their responses when alone, when in the company of females, and when in the company of other males. Unfortunately equipment failure, back injuries, and a shortage of time prevented me from completing my experiment during the field season in which it was scheduled, so we do not yet have the results.

If males really do monitor the amount of cooperation they are getting from their colleagues, and adjust their investment in particular social relationships accordingly, then we would expect males to participate more actively when they are being observed by the other males of their group. I recall one memorable intergroup encounter in which two males "sneaked" downhill away from the combat zone via separate routes, and then bumped into each other in the course of fleeing, while the fight was still continuing on top of the hill. When they saw each other, they looked startled, grimaced and screamed at each other, and then charged back up the hill together to rejoin the battle.

Although I could fairly effectively rule out the hypothesis that capuchin females are bonded with other females because of their need to cooperate in intergroup encounters, it was harder to rule out the hypothesis that males are, at least in part, "hired guns." Regardless of the nature of the benefit males were providing to females—defense of access to resources or defense against infanticide—it remained possible that females actively reward males for providing effective defense. Of course males obtain direct benefits from participating in intergroup encounters, in addition to any that the females might offer, but encouragement from females might conceivably increase the level of participation of subordinate males. I analyzed the data from Abby's group in the first two years to see whether there was any evidence of active encouragement by females of male participation.

Females did little or nothing to encourage males at the outset of an intergroup encounter, or during it: they were too busy running away. But the females' responses to males returning from intergroup battles were quite varied, ranging from threats to ignoring to enthusiastic gargling, sex, and grooming. Females enthusiastically greeted returning males only when they had been victorious or had "tied" with the other group. (I scored an encounter as victory or a tie if the males of Abby's group did not lose any ground to the opposing males before the males of both groups returned to their females.) In most cases, the alpha male was the target of the greeting. However, the copulations we witnessed in the wake of successful intergroup encounters also involved subordinate males with pregnant females.

Perhaps the most puzzling aspect of intergroup encounters is the high frequency of dancing and strange vocalizations that occur in the midst of a confrontation. Males can be highly ambivalent toward one another. Whereas the primary goal of intergroup encounters is clearly to chase the foreign males away, it is obvious that males are intensely curious and excited about foreign males. Although they fear them, they also want to find out more about them. It is exceedingly rare for a male capuchin to spend his entire adult life in one group; typically males transfer every three to four years. Intergroup encounters offer opportunities for males to learn more about their transfer possibilities. Individuals who are their mortal enemies today could be their allies next year, if they play their cards right. Perhaps the dances, which are largely identical to courtship dances, aside from the vocalizations, are a way of negotiating relationships between males of different groups.

Other common behavioral elements in intergroup encounters are branch breaking, piloerection (fluffing of the hair), tooth grinding, bouncing, and urine washing. Open-mouth threats are a way of showing off the size and quality of canine teeth, whereas tooth grinding (often more like tooth gnashing or clacking) produces a surprisingly loud sound that may indicate jaw strength. Branch breaking displays the strength of the arms and legs. Males also often seem to show off by sitting in a dangerous place, such as the middle of a road, during an intergroup confrontation. In fact, Abby's adult son Rico was killed by a car while standing in the road during an intergroup encounter.

**April 20, 2003**     It is around 8:30 in the morning. Two field assistants, Nando Campos and Juanca Ordoñez Jiménez, are following Rambo's group when they hear a chorus of intense vocal threats, some screams, and a wheeze. As they run toward the sounds, someone starts doing a series of alarm calls that sound a bit like snake alarms and a bit like terrestrial predator calls. More voices join in the threatening. When Nando and Juanca reach the site of the commotion, they discover a monkey partially submerged in the river, apparently treading water, near Gizmo and Skippy, a pair of subadult brothers, who are glaring at him. The monkey in the water alternates between alarm calling and doing some roars and undulating screams. Whereas adult females and juveniles of both sexes scream often, it is extraordinarily rare for adult males to scream.

"*Es Jordan!*" calls Juanca. Sure enough, the gaunt adult male of approximately twelve years of age is Jordan, struggling in the water, looking absolutely terrified. We had met Jordan the previous year, when he immigrated into Abby's group with his constant companion Jackson. Jordan was Jackson's chief henchman and had helped Jackson attain alpha rank. But the two of them often made excursions away from Abby's group. This month, Abby's group was falling apart, and its members could be found in various combinations scattered all over the forest. Today, Jordan appears to be completely alone.

Soon most of the Rambo's-group "bad boys"—Tranquilo, Barbell, Rocky,

Badoodie, Gizmo, Skippy, Bart, and Duende (all sons of Pablo)—have assembled and are surrounding Jordan, threatening him, swatting him, and shaking branches at him. Many of the males have penile erections as they glare menacingly and lunge at Jordan. Jordan alarm calls at a rate of about once per second, switching to undulating screams whenever someone lunges at him. He has a fresh, deep gash on his left shoulder. While the attacking coalition is momentarily distracted by a fight elsewhere, Jordan cautiously emerges from the river and attempts to slink away. He is immediately detected, and there is a cacophony of intense vocal threats as the males chase Jordan and back him into a corner in the riverbank.

More monkeys come over to investigate, including alpha male Pablo and alpha female Mezcla. Monkeys come and go, but there are typically five to eight males, mainly subadults, surrounding Jordan at any point. Occasionally the males recruit a female to help them threaten, but the females are not as interested in tormenting Jordan as the males are. On several occasions Jordan makes a break for freedom, but each time it looks like he will escape, several Rambo males chase him down, pounce on him, and wrestle and bite him into submission again. Jordan is so afraid to turn his back on his attackers that he tries sneaking away backward. After this series of attacks has lasted a full hour, Jordan finally does manage to escape, and he returns to Abby's group a week later.

It must be terrifying to be a migrating male caught alone. I had a huge adrenaline rush just listening to the recordings of Jordan's encounter with Rambo's group, despite the fact that I knew it had a happy ending. I had seen Jordan survive many a nasty fight in Abby's group without uttering a squeak, so it was heart wrenching to hear him sound so completely panicked that he screamed and alarm called continuously for an hour. One of the remarkable things about observing an attack like this is the rapidity with which the monkeys change moods. Rambo's group is usually a cheerful, happy-go-lucky group in which, on a typical day, "nothing happens." (Observers of Rambo's group tend to have less exciting monkey gossip to report than observers of other groups.) The same subadult males who were so vicious to Jordan normally spend all day cuddling with their groupmates, playing, and caring for infants. Females readily allow them to take their tiny youngsters for long rides and play with them, and they are usually extremely tender with the infants. But when a foreign male appears on the scene, they switch emotional gears entirely.

I recall another time, when a lone immigrant subadult male tried to migrate into Rambo's group. A native young male of the same age was initially friendly with him and tried to dance, but when the migrant male was spotted later on by Pablo, the alpha male, Pablo threatened him and it was all over. An extraordinarily vicious mobbing ensued, and all of the Rambo's bad boys piled on top of him. At least five of them were chewing on him, biting out chunks of flesh while he screamed and alarmed as Jordan had. I was so shaken up that it was at least an hour before my pulse returned to normal. The males also remain pumped up for a long while after these encounters: their hair stands on end, and they swagger about looking edgy and vigilant.

It is particularly easy for such emotions to get out of hand in a group containing many males, like Rambo's group. Around the time of these events, Rambo's group included sixteen males who were old enough to participate in intergroup encounters, and most of them were close kin. Whereas their social interactions within the group were typically extremely amiable and cooperative, together they formed a terrifying coalition against foreign males. It is no wonder that males tend to migrate with partners, rather than attempting to migrate alone. Being caught without one's allies by a coalition of foreign males is a deadly situation, even if the lone male is doing nothing more offensive than taking a nap. Not all monkeys are as lucky as Jordan was, unfortunately. The following scenario is one that has become all too familiar to us over the past few years.

**February 13, 2002**     I had just left Yellow House, our project's new headquarters, where I had been working with Hannah Gilkenson on a new museum exhibit about the project, and had begun walking across town toward the old house, where I was staying with Joe and Kate. To my puzzlement, I saw the battered old *mono*mobile (the "monkeymobile," our 1977 Land Cruiser with monkeys painted on the side) crossing town in the opposite direction from home. Intrigued, I jogged ahead and flagged it down.

"What's up? You're going the wrong way," I said.

"Oh, *there* you are! We were looking for you," said Brent Pav, who was driving. Brent normally was cheerful and unflappable, but at the moment he looked spooked. I glanced into the backseat and saw that everyone in the car looked tense. "This huge fight happened at 5:30, as we were leaving the monkeys," he said. "So we ran back. Eight or nine Rambo monkeys were on top of this guy, biting him as he lay on his back. He wasn't even trying to fight back; he was just lying there looking subservient and screaming constantly. Every time the males came closer he screamed even louder. They dragged him by the tail, his guts were spilling out, and he finally escaped to the stream. It was awful, he was lying there screaming with his face barely sticking out above the water, and the water was pouring into his mouth while he screamed. It was just awful, I've never seen anything like it! What should I have done?"

"Ugh. You don't know who it was?"

"No," he shrugged miserably. "It was already dark and really hard to tell. Do you want to hear the tape? Mino got it all on tape."

I got in the car and we all rushed back to Yellow House and hovered anxiously around Brent on the porch as he rewound his tape and started to play it. Brent's recorded voice sounded strangely calm and bewildered as he attempted to narrate what he was seeing, and it was almost drowned out by about ten monkey voices threatening and screaming at the top of their lungs. You could tell Brent had been shocked almost speechless by what he saw:

OK, there's a monkey lying on his back, he's surrounded by about six monkeys, or maybe eight . . . that's the one on his back doing all the screaming . . . I can't tell who everyone is, but there are a bunch of juvenile males standing around on the edge of this group, and some larger males are harassing him . . . someone just lunged at Doble from behind while Doble screamed at some other monkey . . . I'll just call him Doble for now, but I don't know who it is . . . Pablo is biting him on the neck and dragging him away . . . that's still Doble screaming. OK, now someone else is coming up and dragging Doble away by the tail about a meter . . . some subadult . . . Moth or Newman, I think . . . Now a bunch of them are hitting him, I'm not sure who, and he's screaming louder . . . OK, now they are backing away and just staring at Doble. It looks like they are scared to attack if no one else is attacking. One of these subadult males has blood on his shoulder and arm. Right now Doble isn't trying to defend himself at all, but he screams louder when anyone gets closer. Now he's standing up and walking toward us. Mino's getting out the recording equipment. Good thinking, Mino! I forgot we had that. The other monkeys are slowing down a bit now that we are this close. I wonder if he thinks they won't follow him if he gets closer to us. He's still coming closer. Oh my God! I think his guts are hanging out. Yeah, that's for sure. This is bad. He just walked right across my boot! The rest of the monkeys are pausing now. Oh no, now Moth is tearing after him with a huge open-mouth threat. The others are following. OK, I gotta run now . . . [we hear Brent running and panting as he follows the monkeys] Doble is in the *quebrada* [the stream]. He is almost completely submerged in the water. Only his face is showing, and he is still screaming. About eight or ten monkeys are surrounding him. They are jumping around on the rocks in and around the stream like they are taunting him. Newman is being especially *bravo* . . . he is bipedal in front of Doble on the rocks. With Doble almost submerged, Newman looks really huge, lording over him. They aren't hitting him right now, but he's still screaming. His screams sound funny, I think he is swallowing water. OK, now Elmo is getting *bravo* and lunging at him a few times. Mino, get your flashlight! We've got to find out who this is.

Brent's narration, through no fault of his own, was not as informative as we would have liked; it had been way too dark to identify most of the monkeys. We all listened in horror to the screams of the victim, which were the worst I had ever heard. It would have been obvious even to an untrained ear that this animal was utterly terrified. We didn't hear any human alarm calls from him, which worried me, as it probably meant the victim was someone we knew well rather than an intruder from an unhabituated group. Field assistants Whitney Meno and Gayle Dower were starting to look sick. The men on the project were pumped up with adrenaline after hearing the tape and wanted to listen to it again and again, to try to glean a bit more information from the monkeys' chaotic vocalizations.

We all agreed that we should abandon following Abby's group the next day so that the entire staff could search for the wounded male in the morning, and that we would start out extra early so that we would have a better chance of finding him. As I left Yellow House to go home for dinner, Mino Fuentes was getting out the Sony Professional Walkman and some speakers so that they could listen to his higher-quality recording of the event. I didn't think I could take any more of that poor monkey's anguished voice, and I left the others arguing about how many times they should have to listen to the tape. Who could it be? I anxiously ran through in my mind all of the migrating males I could think of, trying to picture what each one of them would probably look like now.

The ride to the forest next morning at 4:30 A.M. was extremely tense. Most of us hadn't slept much, and those who had had dreamed about monkeys. We arrived well before dawn. When the monkeys started to wake up, we discovered that Rambo's group was widely dispersed, and so we split up in an effort to flank the sides of the group. Soon I heard three whoops from Hannah, which is the signal to come quick, and I rushed toward the sound. They had found the wounded monkey up in a tree just above the site of the attack. I knew from one glance at Hannah's face that it was really bad news, but all she said was, "I think I know who it is, but look for yourself." At first all I could see of the monkey was a huge coil of bloated intestines, brightly glistening, emerging from his right side. Then, when he turned his head, I knew immediately that it was Kola.

Kola had always been a favorite of mine. He was the son of Wiggy, my favorite of the original Abby's-group females, and, like Wiggy, he had a temperament that was so sweet and gentle that it was hard to believe he was a capuchin. He was nicknamed "the *taxista*" (cabdriver) for his habit of ferrying frightened infants across rivers and roads. Kola had an interesting history: he had emigrated from Abby's group in 1996 and was occasionally seen with his brother Quizzler and another male on the Hacienda Brin D'Amour during the following three years. In 1999, after his mother had died, he rejoined Abby's group for two years as a subordinate male and continued to be remarkably gentle, despite the fact that he was at the age at which most male monkeys are intolerably impulsive and aggressive. A year ago he had again left Abby's group, this time in the company of three of his brothers and cousins, and we had spent a bit of time following them and documenting the remarkably relaxed social relationships in this all-male group. Hannah was the only other member of the current crew who had known Kola. When I turned to her and said, "Kola," she nodded sadly in agreement and gave me a hug.

Oddly, even in his current predicament Kola looked as serene and kind as ever. It was surreal. I walked around the bottom of the tree, examining him from all angles. He had several bare patches of skin from recently healed wounds, but aside from that, and the fact that his guts were hanging out, he seemed to be in excellent shape. He was muscular, had a nice thick coat of hair, and showed no trace of the gauntness common in lone migrant males. He seemed quietly alert. Wild capuchins frequently astound me by recovering rapidly and completely from terrible wounds, but, sadly, there did not seem to be any chance of recovery in this particular case. There was about a foot of swollen intestine hanging out of his side, and Kola was making no attempt to push it back in. Occasionally he made some feeble attempts to catch insects that were feeding on his innards, but he didn't even try to clean out the dirt and plant debris that had penetrated the wound. It looked hopeless. There wasn't anything we could do besides attempt to make his last hours as stress free as possible. After Hannah had taken a few pictures to document Kola's dreadful situation, I sent everyone else off to follow Rambo's group. I didn't want to attract the group's attention to Kola by having a lot of people near him. I knew from previous experience that capuchins are not satisfied to fatally wound a victim; they can be downright vicious with a dying monkey, even when it is obvious that the monkey is now completely harmless. We decided that I should be the one who stayed with Kola, since he knew me best.

After Rambo's group and the rest of the humans had left the vicinity, Kola carefully climbed down from the tree and walked to the stream. He immersed the front half of his body in the water for long periods of time, as if he were feeling feverish, but he did not wash the wound. He drank, but he did not eat anything. As the hours progressed, he began to look more and more tired. He spent most of the day either lying in the stream or lying on rocks. At one point he came quite close to me, sitting on the same rock I was on, and I could see tiny cuts all over his back and limbs. Now that I could see all of these smaller wounds, it was easier to imagine the gang attack of the previous evening. As I sat there peering at his wounds, he gazed up at me with a sad, questioning look in his eyes.

It was undoubtedly the hardest moment I had had in my long years of monkey watching. Had there been anything I could have done to help Kola, I would have faced quite a serious moral dilemma. No matter how hard we try to be detached scientists, we all inevitably form emotional attachments to individual animals when we have known them for years, and it is excruciating to see them suffer. According to the ethics of the scientific profession, we should not interfere. Field researchers who study the behavioral ecology of primates are not supposed to intervene in their subjects' lives (unless, of course, they are compensating for a problem that their presence has caused, for example treating a wound caused by trapping or marking a study subject). We scientists are charged with documenting natural occurrences, and we are not to intervene even when the behaviors we see are disturbing (such as predation and killing by members of the same species). If we were to intervene by giving veterinary care to favorite animals, then the demographic data we collected would be meaningless for understanding population dynamics.

Another problem is that my research group does not have permits to capture wild primates. If I were to capture a wounded animal and take it to a veterinary hospital, where it would most likely die from the stress of capture and treatment, then I would be guilty of poaching a wild animal protected by the Convention on International Trade in Endangered Species (CITES)—quite a serious offense, no matter how good my intentions might have been. And also, in my experience, there is rarely anything that even a well-intentioned and qualified veterinarian can do to help. Wild primates are extremely tough and resilient when left to their own devices, but stressful circumstances (such as being removed from their natural habitat and friends and family, and being handled by unfamiliar humans, which are viewed by capuchins as potential predators) can greatly reduce immune function and reduce an animal's chances of survival. I feel pretty confident in saying that in 99 percent of injury cases, the monkey is better off being left on its own than receiving care from humans. For Kola, however, clearly only prompt surgery and antibiotics could have saved the day, and by the time we had found him it was already several hours too late for that.

So given my circumstances and Kola's, there wasn't anything I could do but wait and be miserable with him. By 4:00 P.M. the end had still not come, and by that time I was so sickened by seeing Kola suffer that I gratefully accepted field assistant Kristen Potter's offer to take the next watch and went to train some new field assistants. She came back in an hour's time to say that it was over, and Brent and Mino offered to bury Kola's body, since it would be less painful for them than for me. (We had started burying monkeys' bodies and marking the sites, so that the bones would be retrievable for genetic analysis or for mounting for educational displays.)

In 2004 the Max Planck Society asked us to create an exhibit to commemorate the founding of the Max Planck Institute for Evolutionary Anthropology, and Kola was exhumed, sent to Leipzig, and reassembled (though some bones had been carried away by other animals). The story of his life accompanies the exhibit, so he has a memorial of sorts.

Unfortunately, Kola's violent death was not an isolated incident. We know of two other males that were definitely killed by other monkeys, and another that was probably killed, between 1998 and 2003.[8] The first case was the killing of Ichabod, the Abby's-group alpha male, by his own groupmates in a "revolution." The other incidents were gang attacks on males from other social groups: the males of Rambo's group killed the alpha male of the neighboring Splinter group, and the Rambo's-group males almost certainly killed a male from another neighboring group as well, during an intergroup encounter (we never found the body; the victim vanished and when last seen had two broken limbs). Sometimes we see gang attacks on migrants that are quite severe, but we cannot be sure whether the victim lived or died.

Aggression is an important part of social life in almost all group-living primates. Some primatologists dispute this claim, pointing out that in every well-studied species the animals spend more time engaged in friendly interactions than aggressive ones, and more time in nonsocial activities, such as foraging and traveling, than in all forms of social behavior combined.[9] Certainly this is true of our capuchins too, but we wonder about its relevance.

To answer certain questions, of course, behavior rates per unit of time are the best measures. For example, if we want to know whether females have friendlier relationships with each other than males have with each other, it makes perfect sense to count the minutes during which the animals are grooming or cuddling with members of their own sex. But neither the evolutionary nor the psychological significance of a behavior pattern can be read directly from its frequency relative to other kinds of behavior. Because death permanently terminates an individual's reproductive career, the most important evolutionary statistics about within-species killing are the percentage of the population that falls victim to it, and how many years of reproduction the victims lose—not the relative frequencies of fighting versus grooming. As for psychological significance, consider that among humans in Los Angeles, for example, most adults spend more time commuting than having sex, yet they are far more likely to think about the latter during the former than vice versa! And even in the most violent neighborhoods of the most dangerous cities, you would almost certainly find that friendly interactions outnumber aggressive ones, yet that would hardly justify dismissing the residents' fears and anxieties as unfounded.

Aggression in animal societies is usually held in check by two considerations. First, escalating a fight to dangerous levels poses some risk even to the stronger combatant.[10] A victim forced to fight to the death may inflict damage on his attacker even if he ultimately loses the conflict. Second, as discussed in Chapter 5, aggression is restrained by individuals' reliance on each other for mutual protection and other services, and this pressures them to repair damaged social relationships.[11] Yet relationship value can vary, from a high value all the way down to zero. Individuals of different social groups are likely to be of little use to each other. Perhaps in intergroup conflicts it is only the animals' fear of harm to themselves that prevents fighting to the death. Under what circumstances could attackers kill victims with impunity? One straightforward answer is that attacks by groups on lone victims entail low risk to the attacker.

Accounts of violence among nonhuman primates draw more public interest than almost any other animal behavior research, and it isn't difficult to figure out why. Most Westerners believe that apes' and monkeys' behavior is fraught with moral lessons for humanity, although people's philosophical starting assumptions about this question are diverse and incompatible. Do nonhuman primates give us a view of the Hobbesian, instinct-ruled state of nature from which civilization has redeemed us? Or do they reveal the Rousseauian harmony that civilization has despoiled? Or do they show us, as social Darwinists would have it, ferocious competition as an engine of progress, and our own "natural" inclinations that we suppress at our peril?

First, in cutting through this philosophical knot, it is important to recognize that the equation of "natural" with "morally desirable" (known as "the naturalistic fallacy") is utterly unjustified. Natural selection is an amoral accountant, counting births and deaths with no consideration for fairness, happiness, individual rights, or any other value that most of us would consider important to ethical judgment. No one would argue on behalf of the smallpox virus just because it evolved by natural selection. And many of our most cherished ideals and practices, for example, charity toward strangers, are partly products of "natural" processes probably unique to humans, such as cumulative cultural (as distinct from genetic) evolution.[12] This should not be taken to imply that biologists' findings are irrelevant to moral deliberation. Any ethical theory should be *based on* the best available scientific knowledge, but any valid function relating "evolved" to "moral" will be far more complex than a simple "equals" sign.

Second, capuchin and chimpanzee behavior doesn't show us our "real" selves any more than human behavior shows what capuchins and chimpanzees are "really" like. Each species has its own set of adaptations and predispositions, shaped by natural selection during its own evolutionary history. It is true that closely related species tend to share behavior patterns because of recent common ancestry, but this can't be assumed. There are many instances of dramatic behavioral divergence between closely related species. The much-publicized example of chimpanzees and bonobos is a case in point.[13]

Finally, observations of extreme violence in any particular nonhuman species do not imply that such behavior is "hard-wired" into those animals. Eastern chimpanzees, the subspecies first studied in the wild by Jane Goodall more than forty years ago, have become infamous for their fatal gang attacks on members of neighboring communities, whereas the western chimpanzees studied by Christophe Boesch and others have carried out no such raids in more than twenty-five years of observation.[14]

Does this mean that Kola's death and Jordan's narrow escape are irrelevant to understanding human violence? No, but their relevance is rather subtle. Comparative biology tells us that when distantly related species have evolved the same traits independently, we should suspect the operation of common selective pressures or adaptive challenges. Dolphins look like fish, for example, because that body shape is most efficient for traveling through water. In certain circumstances, humans, chimpanzees, and white-faced capuchin monkeys engage in escalated coalitional aggression, and as a result, all three species have higher within-species killing rates than other primates. This observation raises two linked questions. First, what other common features of these species or the environments in which they evolved, or both, might account for this similarity? Second, what circumstances trigger lethal coalitional aggression in the three species?

Joe had worked with Richard Wrangham to try to answer the first question in 1991, before our observations of capuchin lethal aggression.[15] They pointed out that chimpanzees, unlike almost all other primates, live in social groups called communities, which rarely assemble in the same place but are usually split up into a large number of traveling parties of shifting size and composition. This fission-fusion social organization creates situations in which large parties sometimes encounter lone individuals from neighboring communities. Even if the benefits of eliminating a neighbor in this situation are small, the costs to the attackers are likely to be negligible because of their numerical advantage. In chimpanzees, males stay in the community of their birth whereas most females migrate at puberty, and thus males of the same community are more closely bonded than females. So it's no surprise that males, but not females, carry out cooperative raids into neighboring communities' territories.

This "imbalance of power" hypothesis, as Richard and Joe called their idea, has been given a boost by two recent findings. First, experiments using playbacks of vocalizations have shown that male chimps are much braver about approaching males of another community when they've been led to believe that they have a numerical advantage.[16] Second, male-only raids have been observed in one of the only other primate species that shows fission-fusion social organization and male bonding, the red spider monkey.[17] The imbalance-of-power hypothesis may also explain the lack of lethal raiding in western chimps, which generally travel in larger parties, reducing their vulnerability to attack, compared with eastern chimps. Humans, of course, also travel in groups of varying sizes, and ambushes of lone individuals are the preferred tactic of warfare in small-scale human societies.

However, our observations of lethal coalitional aggression in capuchins pose a problem for the imbalance-of-power hypothesis, for the simple reason that capuchins primarily travel in cohesive troops rather than in small, shifting parties. Although some lethal or near-lethal attacks occur when a group finds a lone migrant, others occur in a within-group context or during intergroup encounters. They have never occurred during raiding parties such as those described for chimpanzees. Thus we have been led to reconsider an alternative hypothesis that Joe and Richard had rejected: that the cooperative lethal aggression is facilitated by advanced cognitive abilities.

John Tooby and Leda Cosmides have proposed that successful coalitions require enforced risk taking: each participant needs to be confident that his comrades won't leave him facing the enemy alone.[18] The only way to enforce risk is to punish deserters, and such social contracts are possible only in animals with sophisticated social-cognitive machinery. White-faced capuchins do have triadic awareness (Chapter 2), and it's possible that they punish companions who fail to take their "fair share" of risks, but triadic awareness has also been documented in some other primate species in which lethal coalitional aggression has not been observed.

Although rates of violence vary enormously among human societies, coalitional aggression (especially by males) is ubiquitous in our species, even in the hunter-gatherer societies that are the closest existing parallel to the setting in which human psychology evolved.[19,20] Social psychologists have shown that people are highly prone to within-group solidarity and between-group enmity, even when the "groups" are formed using bogus criteria,[21] and coalitional favoritism may be more powerful even than tendencies toward ethnic or racial categorization.[22] Obviously, humans can be motivated to go to war or to lynch their neighbors by ideological beliefs that have no parallel in nonhuman animals. And warfare in the most complex human societies is an intricate social, political, and economic institution. Nevertheless, even the most coldly calculating politicians must make use of their foot soldiers' emotional responses and cognitive biases to motivate them. Understanding the evolution of this mental equipment, and using this understanding to devise ways to minimize human violence, will be greatly aided by comparative studies of coalitional aggression in nonhuman primates and other social mammals.

# CHAPTER 10

# Miffin, Nobu, and Abby:
## Capuchin Mothers, Infants, and Babysitters

**July 20, 1995**    At about 8:00 A.M. I come across Vandal lying on her side on a low branch. I have been keeping a close eye on her all week since my arrival, because she is pregnant for the first time and I am curious to see how this female, whom I have known as a particularly whiny and neurotic adolescent, will cope with parenthood for the first time. As I take a closer look, I realize to my astonishment that she has given birth within the past few minutes. I had seen her just an hour or two before, looking extremely pregnant, and now she already has a tiny, wet, black baby on her back. I am surprised to see this, since virtually all births occur at night.

I had always imagined that a monkey giving birth for the first time must be completely astounded and delighted to find, after the ordeal of birth, that she is now in possession of a baby of her very own. Most adolescent females are obsessed with the babies of other females, and Vandal had been no exception. However, Vandal does not seem particularly shocked to have suddenly given birth, nor does she seem particularly interested in the baby. She gets up, saunters over to Paul Bunyan, the current alpha male, and does a squeak threat in my direction, causing Paul to headflag to her and glare at me. Paul doesn't seem any more interested in the new arrival than Vandal is, and the two of them wander off in separate directions.

The baby, whom I immediately christen Mischief, lies asleep on Vandal's back, slung over her shoulders. His tiny hands clutch her neck fur, and his black face is scrunched up into a mass of wrinkles as if he is concentrating very hard on sleeping. Vandal rummages through some leaves and ingests a few insects before sitting down again. The baby wakes up and squirms its way off her shoulder and down to her abdomen, where he begins rooting around on her chest, looking for the nipple. Vandal licks the baby and then turns her attention to her vulva. She begins to pull out the afterbirth and spends quite a long time devouring it and licking blood from herself. In the meantime, the baby finally locates her nipple, hidden away in her armpit, and begins to nurse for the first time. A juvenile strolls by and also ignores the new infant, and Vandal begins to forage on fruit.


Vandal's response to her newborn baby is typical of capuchin mothers. Despite the infant's rather large head size, the birthing process seems to be fairly easy for the female (particularly compared with human births), and after giving the infant a perfunctory glance, it's back to business as usual: foraging, schmoozing with the alpha male and adult females, and caring for older offspring. Capuchins do not take an interest in infants until they are several days old and are starting to become more active. A relatively large fraction of brain growth occurs after birth in capuchins, compared with all other primate species except chimpanzees and humans.[1] In other words, the ratio of the newborn capuchin's brain size to the adult capuchin's brain size is far lower than is the case for other primate species. This may explain why they seem so helpless and undeveloped at birth, and do little besides sleep and nurse. This pattern of slow growth continues: capuchins have a longer juvenile phase and overall lifespan than would be expected for their body size.

A newborn capuchin has a black or grey face and orients itself across the mother's shoulders, perpendicular to the spine. Since the baby mainly sleeps, it is hard to notice that it is there; it looks from a distance like the mother just has broad shoulders. This arrangement is probably useful for avoiding detection by avian predators. But it is not long before the newborn's face starts to turn white, making its presence more obvious, and it starts to move about and take an interest in life. At this point, the rest of the group becomes obsessively curious about the baby.

One of the best-documented first days of life we have in our records is that of Mowgli, the daughter of Miffin, a feisty, high-ranking female from Pelon group. Since Mowgli was one of only three focal infants in that group at the time, Nick Parker and Mino Fuentes were able to follow her for three and a half hours in her first day of life. Mowgli, like most newborns, spent at least half of her time sleeping and moved from her mother's shoulders only twice, to squirm into nursing position. Her mother responded to her daughter's arrival in much the same way that she responded to any sort of social excitement: by grabbing random bystanders and making them have sex with her. When adult females are in an advanced stage of pregnancy or early in the postpartum phase, they frequently engage in high rates of nonconceptive sex. But Miffin is more extreme than your typical female. She is a favorite focal animal because she is unbelievably meddlesome and melodramatic, and uses force where most monkeys would use tact. In this three-and-a-half-hour period, Miffin sexually mounted nine different individuals at least once each: three adult males, one adult female, four adolescent males, and one juvenile female. Fonz, however, she treated with more deference. Fonz, Miffin's father and the alpha male, was not the father of Mowgli. Mowgli's father was Champignon, an adult male who had no interaction with the baby at all during her first day of life. But Miffin spent quite a bit of time deferentially following Fonz, grooming him, and gargling to him (at a rate of nine times per hour—sixty times the normal rate for a female without an infant).

Mowgli received very little attention from her new groupmates, especially considering how flamboyantly her mother behaved. Miffin vocalized at a rate of at least sixty-seven times per hour that morning, in addition to her conspicuous sexual antics, so no one could have failed to notice the new baby's arrival. The two youngest infants in the group, Camden and Shylock, greeted Mowgli perfunctorily, and one adult male inspected her as well. Mowgli's two-year-old cousin Capulet was most eager to meet her, inspecting her three times and grooming her once. Miffin's younger sister Calabaza ignored Mowgli, even when Miffin

Juvenile female Broma inspects the dying infant Omni, who is tucked between Abby's leg and torso. Squint is in the background. Photo: Susan Perry.

rudely nursed from Calabaza while her daughter Capulet suckled from the other nipple. Miffin groomed her baby five times and chewed on the umbilical cord a bit, but other than that, she ignored her infant in favor of foraging and socializing.

A capuchin's early motherhood experience is entirely different from that of a human mother. From birth, a normal capuchin infant is capable of clutching onto its mother's fur twenty-four hours a day, even while sleeping, leaving the mother's hands entirely free for foraging and traveling. The infant can navigate from its mother's shoulders to her belly on its own and is capable of finding a nipple without help, though it sometimes takes a few minutes. So really the only thing the mother has to do is move her arm out of the way a bit (capuchins' nipples are in their armpits) when the infant needs to nurse. If necessary, she can eat and travel and even leap from tree to tree while her infant is nursing.

I have to say I was highly envious of all the capuchin mothers I had known when my own daughter was born. They made mothering look so easy. Human babies, of course, are incapable of clinging to their mother, and even with the help of various carrying contraptions it seemed that I rarely had even one hand free, never mind two, during my daughter's first year of life. And nursing a newborn human typically requires the mother's full concentration and two hands. I found myself thinking that if it weren't for biparental care, I wouldn't have been able even to feed myself, let alone do any academic work, during Kate's first year of life.

One of the most striking aspects of the capuchin mother-infant relationship, from a human perspective, is the complete lack of *en face* interaction. Human mothers spend vast amounts of time cradling their infants in their arms, gazing into their eyes, and deliberately engaging their attention by vocalizing and making interesting facial expressions. Capuchins do none of this. Usually the infant is on the mother's back, and when it is ventral, nursing, the mother may groom it or occasionally lick it, but she will not look into its eyes. It is also rare for capuchin mothers to direct vocalizations toward their infants when they are in contact with them. Mothers will sometimes utter social peeps when their infants are near, though it is not clear whether they are directed toward their own infants or to other individuals sitting next to them. Most mothers do not even trill to their own infants, particularly when they are in contact with them.

Although capuchin moms are quite perfunctory in their care of normal infants, they can really rise to the occasion and show astounding devotion when an infant has a birth defect that renders it incapable of caring for itself. For example, in 1993 Abby gave birth to Omni, a baby born with an umbilical hernia (see Chapter 6). The birth took place in the morning, and Paul Bunyan and Squint, Abby's constant female companion, were beside her for the delivery. Squint observed intently as Abby pulled the baby out. The whole delivery process took under thirty minutes. Omni received far more interest than normal babies receive. The trouble began immediately, when Abby bit through the cord, or what she (and we) thought was the cord. Intent on cleaning her baby up, Abby kept nibbling away at the cord, and over the next few hours it slowly dawned on us that this "cord" kept growing longer rather than shrinking: this well-intentioned mother was eating her baby's intestines.

Unlike most infants, Omni was incapable of holding on, so Abby had to clutch him to her ventrum as she moved about, sometimes walking bipedally. Abby ate the placenta, sharing a bit of it with her daughter Diablita, and occasionally inspected and licked Omni. Abby took great care when she held Omni, cradling him gently in one arm rather than casually dragging him about. When she sat down, she curled her tail around his body to protect him. Omni continued to be listless and failed to nurse, though he occasionally squirmed (especially when Abby chewed on his "cord"). Often he faced away from her ventrum, instead of toward it the way a normal infant would. Abby's adult daughters noticed the trouble and came over, gingerly touching Omni and sniffing him. Omni continued to look limp, and his hands were relaxed most of the time, though he occasionally gripped for a few minutes.

As the group moved on, Abby began to experiment with new techniques for transporting the infant. Climbing tree trunks was particularly challenging: at some points she needed two hands to climb, and so she would gently press Omni between her belly and the tree, pull herself up a few inches, and then reposition the baby higher up. As the day progressed and she became more desperate, she became a bit less gentle in her handling of the baby, sometimes holding him upside down with just one hand as she made a flying leap to another tree. But she never abandoned the baby even for a second. Twice she dropped Omni, but she immediately alarm called and ran down to retrieve him. Over the course of the next three days, before Omni died, Abby and Omni received an inordinate amount of attention from the rest of the group. Abby tried to shield the baby from excessive poking and handling, but her older offspring were increasingly fascinated by their new sibling's bizarre behavior.

Abby's daughter Maní had a similar experience on April 26, 2004, when she gave birth to a highly deformed stillborn infant we called Pobrecito. This was the most peculiar infant we had ever seen. In many ways the infant looked premature: there was hardly had any hair on the body, for example, aside from a ridge of hair along the spine. But the bones were immense, as if the rest of the baby's development had not kept pace with the skeletal growth. Also, the head was covered with an excessive amount of hair, and the facial hair had already turned white, whereas infants are normally born with gray/black facial hair. There were no wounds on the body at all when Juanca and Heidi arrived on the scene, so we had no reason to suspect that this infant had been the victim of infanticide.

Maní was carrying the dead infant everywhere she went, though she was not treating the body with the same care that Abby had given to Omni. Sometimes she grasped Pobrecito around the waist, and other times she held on to the tail or another appendage, letting the body dangle as she locomoted. When she needed to eat or groom someone, she set the body down on a branch, usually continuing to grasp it with a foot. She did, however, occasionally groom the body of her dead baby. Her son Marañon inspected his dead sibling frequently.

Around 9:00 A.M. flies started buzzing around the body, and this irritated Maní, who kept trying to catch the flies. Maní became preoccupied with the baby's anal region and repeatedly licked and sucked on it, as if ingesting fluids emitted from the body. Although she was concerned with keeping the anogenital region clean, and with keeping the body free of insects, she did not treat the body as if it were alive in other respects. For example, whenever she got a drink from the river, she allowed the body to be completely submerged in the water. Maní carried the body all day and slept with it that night.

The following morning when I arrived with Heidi, Maní was still carrying the body. The corpse was starting to wither and dehydrate, looking now like it was all head with just a stick body connected to it. Around 6:00 A.M., carnivorous wasps discovered the body and began to feed. These wasps are some of the peskiest creatures in the forest: they sting with one end and chew with the other, they live in colonies of hundreds (at least), and they will pursue you for a kilometer or more. Maní was kept busy constantly swatting these insects away. She continued to groom, lick, and suck at the body throughout the morning. Maní also seemed worried that Duende, one of the adult males, would take the body from her. Occasionally, however, she would set the body down on a branch to forage for a bit.

Various group members came by to inspect the body and threaten it. In one of the stranger moments of the morning, Maní's sister and daughter formed a coalition against the dead body. And then Maní backed into their overlord gesture and joined them in threatening her own dead baby. She also made agitated "scary-food" peeps while looking at the insect-covered corpse. As the fly and wasp infestation grew worse and worse, she began making these scary-food peeps incessantly. The monkeys normally use this vocalization when they are inspecting or foraging on something that is edible yet potentially dangerous—like stinging caterpillars, wasps, or small rattlesnakes. They also make this sound when looking at wounds or insect-covered monkey feces. This vocalization invariably attracts the attention of nearby monkeys, who come to inspect the object of interest and often join in coalitions against it. In this case, Duende and Opie came over to inspect Pobrecito and join Maní in a chorus of scary-food peeps.

It is difficult to eat *Sloanea* fruits while holding a dead baby, and eventually Maní left the remains propped carefully in the crook of some branches while she moved a few meters away to forage. "Oh no," I groaned, "Don't leave it there!" The crook was about twenty-five meters up, and there was no way we would be able to collect the body to get a tissue specimen for genetic analysis. Fortunately, a few minutes later she returned and retrieved it.

By 7:45 A.M., the baby's torso had shriveled to the width of a fat caterpillar, and it was completely covered with wasps, despite Maní's frantic attempts to keep the body free of insects. She kept up a constant stream of agitated peeps and alarm calls, and kept lunging at the flies and wasps. The wasps kept stinging her, and she scratched and twitched constantly, looking like a nervous wreck. But she was not willing to give up the body. However, when she accidentally dropped it, Elmo and Maní herself formed an overlord against the body before she went down to retrieve it. The adolescent females Fishy and Cookie ran down to inspect the body, but Maní chased them away. A few minutes later, Marañon inspected the baby, first touching it gently and then threatening it and knocking it out of the tree. Maní again rushed down to the body, but before she picked it up, she paused to stare at it. She alarm called, swatted at the wasps, and did another long sequence of scary-food peeps before picking up the body, seeming reluctant this time.

"This is so stressful to watch! I just feel so sorry for her," said Heidi.

"I know. This is horrible. I'm sure she'll feel better when it is gone and she doesn't have to deal with these constant wasp stings. Surely she'll abandon it soon. I hope so anyway . . . it can't be healthy for her to be sucking on this rotting body," I said. "But I understand why she can't just abandon it." Maní's previous baby had been killed in the infanticide spree of 2003, so it seemed particularly hard luck that her next infant was stillborn.

Around 9:15, Maní left the body on the ground and went to forage. When she didn't return for a while, we picked up the body, put it in a plastic bag, and hid it from view, so that we could get a tissue sample from it later. Maní eventually came back to look for it. She scanned the ground for a bit, looking perplexed. Then she and Elmo formed a coalition and threatened toward the patch of ground where the body had been. But after just a few more seconds of searching, she relaxed and resumed foraging. It was as if she had felt compelled to protect her baby so long as the body was in her view, but once it was out of sight, she felt free to resume her normal life. When we were convinced that Maní no longer wanted the dead body, I sent Heidi back to Bagaces with it to preserve some of the tissue for genetic analysis.

When infants are developing normally, caring for them seems to cause their mothers very little stress. Infants can even nurse from a dorsal position, since the mother's nipples are in her armpits and can be reached from the back. Infants tend to nurse until they are about two years old—that is, until the next infant comes along—and infants whose mothers have an extended interbirth interval will nurse even longer. During their first few months of life, infants nurse one to three times an hour, and even when they are a year old they nurse once every hour or two. During the first three months of life they take essentially no solid food, and they do very little foraging on their own during the first six to nine months. In most cases, nursing starts to taper off to once every second or third hour at the end of the second year of life, though an infant will continue to nurse a couple of times per day even when three years old, if his or her mother has not yet had another baby. There is a huge amount of inter-individual variability in nursing rates even among infants who are the same age. Perhaps because of their dietary diversity and the importance of extractive foraging in the capuchin diet, capuchins take many years to hone their foraging skills. Milk therefore seems to remain an important part of their diet for quite a long time—certainly well into the second year, if not longer. Insects are the main source of protein for adults, and they are hard both to capture and to extract. Juveniles continue to make dramatic progress in learning to extract insects from sticks even in their third year of life.

Once they become alert and active, around two weeks of age, capuchin infants really take charge of their own development and their interactions with their mothers. Mothers are not particularly restrictive of their infants' movements; infants are given free rein to explore their surroundings. During the first three months of life, 1 to 2 percent of their time is spent on the back of an alloparent (a caregiver who is not their mother). At four to six weeks of age, infants start to show greater interest in the members of their social group who come to inspect them, and they often crawl onto the backs of these visitors. Carrying by alloparents peaks at this age, when excited alloparents carry the infants 2 to 9 percent of the time, often taking them away from their mothers on long excursions. Once infants reach seven months of age, when they no longer need to be carried so much, they are typically carried by alloparents for less than 2 percent of their time, about the same amount of time their mothers carry them.

Alloparenting (also called babysitting and "aunting") is fairly common among primates, though capuchins do it more than most primates. Why primates engage in alloparenting has been a topic of heated theoretical debate.[2] After all, caring for an infant that is not one's own seems on the surface to be a waste of time and energy, since it detracts from the caregiver's foraging time and from her ability to invest in her own offspring. On the other hand, many people not trained in behavioral ecology would not find this question mysterious at all. Obviously, everyone is interested in babies because babies are cute.

However, just as evolutionary psychologist Donald Symons has argued regarding the study of sexual attractiveness ("Beauty is in the adaptations of the beholder"), cuteness is not an irreducible property.[3] The small size and relatively large heads and eyes of baby mammals sometimes stimulate friendly, nurturant responses in their mothers' companions and other bystanders, but at other times they elicit murderous aggression. And the attitude toward infants is not predictable just by the sex of the bystander. In common marmosets, a monkey species in which the dominant female usually monopolizes breeding, it is the alpha fe-

Cassie (right) babysits an infant male, Outlaw, with whom he has developed a close relationship. Photo: Susan Perry.

male who often kills the infants of subordinates who reproduce.[4] So we need to ask how the common *perception* of others' infants as cute, and the ensuing desire to touch and carry them, is beneficial in evolutionary terms—that is, in promoting survival and reproduction. There is no guarantee that a single explanation will apply across all species, or even across different classes of individuals (males versus females, juveniles versus adults) within a species.

To understand one hypothesis, first advanced more than twenty-five years ago, consider the analogy of a grazing animal whose preferred food species is most nutritious near cliff edges.[5] How close to the cliff should our grazer forage to maximize her chance of surviving and reproducing? The costs of the two potential kinds of mistakes (nutritional deficiency versus falling of a cliff) are very different, so it pays to err on the side of caution. The evolutionary trade-offs surrounding females' fondness for infants represent a similar dilemma. Being too attracted to babies could lead to some wasted time and energy, but, as primatologist Joan Silk puts it, "too little and you might leave your baby on a street corner." Thus, in this hypothesis, attraction to other females' infants is merely a by-product of selection for devotion to one's own infants. It predicts that the most active allomothers will be females currently lacking small infants of their own.

A second idea with a long history in primatology is that allomothering serves as practice for real motherhood.[6] Primate maternal skills, though based on innate predispositions, must be learned, and first-time moms are often remarkably clumsy and inept. This is one reason why, in a wide range of species, females' first infants are more likely than subsequent offspring to die in infancy. If this motherhood-training hypothesis is correct, then we expect, of course, for allomothering to be a particular preoccupation of adolescent females, and we expect the first-time moms with more allomothering experience to be less likely to lose their infants. Certainly adolescent capuchins are keen on handling babies, but we do not yet know whether this early experience is helpful to them once they have infants of their own. And since our findings show adult females, rather than adolescents, are the most active infant handlers, practice certainly could not be the only explanation for the phenomenon of alloparenting in capuchins, although it could be one contributing factor.

As the evolutionary theories of kin selection and reciprocal altruism became influential during the 1970s, both were invoked as possible explanations for allomothering. Perhaps females preferentially babysit their close relatives' offspring.[7] Or maybe females trade allomothering services.[8] In either case, infants should benefit, in terms of survival chances, from frequent allomothering (or, in a slightly more complex argument, mothers should benefit by being able to forage unencumbered while someone else carries their infant). Besides supplementing maternal care, allomothering could provide a kind of "life insurance policy" by creating bonds between infants and potential adoptive mothers who could care for them if they were orphaned.

However, as primatologists contemplating the darker aspects of sociobiological theory paid closer attention to what actually happens when female monkeys and apes touch or carry one another's infants, some began to doubt whether such benign terms as *allomothering* and *babysitting* were even warranted. Sometimes the interactions are rough, it was noted. Occasionally an infant is "kidnapped" (kept away from its mother when she obviously wants it back) or even "aunted to death": carried by a nonlactating female (of higher dominance rank than its mother) until it dies of dehydration. Joan Silk and Sam Wasser independently proposed that allomothering is actually a form of female-female reproductive competition, and functions to decrease the survival chances of rivals' infants.[9,10] Primatologists began using the more neutral term *infant handling* to describe the phenomenon. If infant handling is indeed a form of harassment, then handlers should usually outrank the mothers of the infants they handle, and the handling should do infants more harm than good.

In general, the types of infant handling we have witnessed among capuchins are benign. On those rare occasions in which infant handling crossed the line to become something that might be considered abusive, the caregivers were adolescent males. Such males have sometimes exposed an infant to risks that its mother might have avoided, or abandoned the infant when it became annoying. We have seen a few extraordinary cases of abuse, when a young male tried to throw an infant off his back, swatting, nipping, and rolling over on it, when it refused to dismount. These situations tend to occur when the infant becomes hungry and starts to coo for its mother, refusing to be pacified by the milkless male, who does not seem to understand why the infant is so distressed. If a male abandons an infant far from other group members, the infant certainly runs the risk of becoming lost, but we have never observed kidnappings like those commonly reported in macaques, in which a nonlactating monkey steals an infant and refuses to return it to the mother when requested.

Most capuchin infant handling, particularly early in life, resembles this vignette recorded in our field notes:

April 26, 1997. Nanny is resting while her seven-day-old son Butler sleeps on her back, his tiny hands gripping the hair on her sides. Squint, whose youngest offspring is two years old, approaches the pair, places her face just a few centimeters from Butler's, trills to him, and then begins a rapid sequence of soft, low-pitched, staccato noises. This "machine-gun vocalization," as we call it, is given only when inspecting another monkey's baby. A few seconds later, Squint reaches out and gently lifts Butler's tail while trilling again. A minute later, Butler crawls around to Nanny's front and begins nursing.

Alloparents seem obsessed with inspecting the genital area of infants and will do it repeatedly, for weeks on end. We joke that they find it as hard as we do to determine the sex of newborns because of the unusual pseudo-masculine form of the young female's genitalia, but we really are not sure why they do this. Whereas this constant handling may be annoying, disturbing the infant's sleeping and nursing activities, it certainly doesn't qualify as abusive. If mothers viewed the handling as abusive, we would expect them to show a great deal of concern when their infants are handled, and we see very little of that.

In a study that Joe did of the social lives of infants during their first three months, mothers of all dominance ranks were permissive as their babies began moving off them to explore their surroundings.[11] In response to their infants' straying beyond one body length away (about sixteen inches), mothers restrained them (such as by grabbing an ankle) only 3 percent of the time. And mothers were not aggressive toward females who sought to handle their infants. On only five occasions in more than a hundred hours of focal observations did a mother threaten or lunge at a would-be handler. Sometimes a mother with a baby on her back would continue resting, eyes closed, while an admirer nuzzled and twittered to her infant.

This general pattern holds true in my current study of infant development, conducted from 2001 through 2006: the mothers I observed typically showed little or no concern when their infants rode away on the backs of other individuals, even when those caregivers were boisterous, irresponsible, risk-taking adolescent males who whisked them into the center of wild play-wrestling bouts. But there was extensive variation in the level of maternal concern and control over these situations. For example, Una and Cupie, the mothers of Ugali and Chaos, never once retrieved their infants from alloparents in their first eighteen months of life, despite the fact that they were often being carried by nonmothers. In contrast, the meddlesome and domineering Miffin retrieved her infant Mowgli from the care of a wide variety of alloparents of all age, sex, and rank classes, at a rate of one retrieval every two hours during the first three months of life, and one every four hours when Mowgli was four to nine months old.

Another alloparenting hypothesis, which we will term the schmoozing hypothesis, states that infant handling is a way for the handler to curry favor with the infant's mother, an action that may pay off later in the form of coalitional support.[8] Like the kin-selection and reciprocal-altruism hypotheses, this assumes that handling benefits infants or mothers, or both—if it didn't, why would it improve the relationship between mother and handler? Unlike the harassment hypothesis, this theory predicts that handlers should usually be ranked *lower* than the mothers of the infants they handle.

Dario Maestripieri, who has studied infant development in several macaque species, has pointed out that infant handling differs among Old World monkey species in ways that make sense in light of other ways that these species differ.[2] In species such as rhesus and Japanese macaques, with strict female dominance hierarchies and high levels of female-female competition over food, infant handling is usually rougher, and females (especially low-rankers) are more protective and restrictive of their infants, than in species with less competition and more relaxed dominance relationships, such as Hanuman langurs. Maestripieri has urged primatologists to carefully distinguish benign from abusive infant handling to see how generally this link holds.

To test some of these hypotheses, Joe looked at the social relationships between each mom and the females that frequently handled her newborn infant (zero to three months old), and compared them with the relationships between the moms and less-frequent handlers. Remember that the harassment hypothesis predicts that handlers will usually outrank mothers, whereas the schmoozing hypothesis predicts the opposite. Neither pattern held for our data sets; dominance relationships were unrelated to frequency of infant handling. According to the reciprocity hypothesis, comparing pairs of females should reveal a correlation between the rates at which each female handles the other's infant. (For example, if Nanny handles Tattle's infant frequently, then Tattle will handle Nanny's infant frequently, and if Wiggy handles Squint's infant rarely, then Squint will handle Wiggy's infant rarely.) We found no such statistical relationship.

Last, we examined whether infant handling in capuchins could be explained by kin-selected altruism. We didn't know the makeup of the monkeys' family trees until 2004, after laboratory analysis was completed on DNA extracted from our collection of fecal samples. At that point it became clear that the adult females of Abby's group comprised a single family with Abby as the matriarch, which meant that up until 1999, when the group grew to include eight adult females, there wasn't enough variation in relatedness among female-female pairs to find much effect of kinship on social behavior. In any case, remember that for infant handling to qualify as altruism (whether reciprocal or kin selected), it must benefit infants. The nuzzling, gentle touching, and machine-gun vocalizations by alloparents might be friendly, but it was hard to see how they could enhance infant survival.


The two most noteworthy benefits to capuchins who receive alloparental care seem to be nursing and opportunities for social learning. Capuchins are remarkable in the extent to which they share milk with others' infants.[12] Allonursing begins when the babies are just a few weeks old and continues for several years, even after the infant has been weaned by its mother. When the babies are four to six months old, 16 percent of their nursing bouts are routinely with females other than the mother. Because capuchins have such wide inter-individual spacing during foraging and travel, it is common for an infant to lose track of his mother for several hours, and during these periods it is tremendously helpful for him to be able to nurse from someone else. By the time the infants are four months old, they are spending only 20 to 40 percent of their time with their mothers and are making long excursions away from them. Many young infants we have known have lost their mothers for at least a day or two during their first year of life, and such a situation could lead to dehydration and death if it weren't for milk from allomothers.

Reports of allonursing in another species of capuchin, *Cebus olivaceus*, indicate that allonursing is often parasitic, with juveniles demanding milk from females whose mothers are subordinate to their own.[13] This is not the case in *C. capucinus*. Although lower-ranking females do tend to provide more allonursing services than do higher-ranking females, there is no correspondence between the rank of the infant (or the infant's mother) and the rank of the milk-providing female. Nor is there a tight reciprocal relationship between females regarding the provision of allonursing services. It is therefore a bit of a puzzle why lactating females so cheerfully provide milk to others' infants. I have rarely seen a female deny milk to an infant, even when she already has her own infant nursing on her other nipple. Sometimes females will even invite nursing by nonoffspring, by lifting their arm to expose the nipple as the infant approaches.

Milk is somewhat costly to produce, and mothers of most species jealously guard against investing in offspring that are not their own. But the cost of giving milk to another's infant is small compared with the benefit of having the favor returned at a time when the mother and her own infant are separated long enough for the infant to dehydrate. In addition, because the females in a group are quite closely related to one another through both the maternal and paternal lines (during times of long alpha-male tenures), kin selection should facilitate the spread of allonursing tolerance.

We observed one case of an infant orphaned at the age of eight months—that is, at the age at which most infants are nursing at least once an hour and sometimes as often as three and a half times per hour. Nobu lost his mother to a poacher. He spent the entire day after his mother's death staying near her body, cooing and trilling piteously, and nursing futilely from her dead body. We didn't think he stood a chance of surviving, as the rest of the group had moved on. The next day, however, he was reunited with the group, and his cousin Celeste let him nurse. His older brother Till gave him rides and hovered near him for the rest of the day.

Nobu was obviously deeply depressed in the weeks following his mother's death. Whereas he had formerly spent only 7 percent of his time alone, he now spent 32 percent of his time out of view of other monkeys, and he declined to join in the rough-and-tumble play that he had enjoyed previously, dropping to about 2 percent of his former play rate.

Although he was receiving milk from his sister Opie, his cousin Celeste, and his many aunts (in fact, from all females in the group), he now had to spend far more time foraging for his own food, and we were impressed by how quickly he matured. He rapidly became the most efficient forager in his cohort. By the time he reached twelve months of age, he was five times as efficient at extracting insects from sticks as the other yearlings were. However, he needed to spend twice as much time foraging as he had previously, which meant that his social time was reduced to half of what it was before his mother's death. The field assistants were all completely charmed by Nobu's bravery and the accelerated development of his foraging skills. Nevertheless, despite all his hard work and the milk handouts from his female relatives, his growth remained stunted. Nobu's primary attachment figure, once his mother was gone, was Abby, his grandmother. Abby was quite old by this time and suffering from arthritis and rather dire dental problems. She spent most of her time on the ground, and this meant that Nobu did too. They were quite a sight, striding across the savanna together looking for acorns and grasshoppers while the rest of the group stayed in the treetops along the riverbanks. We worried that Nobu would not have much chance at a normal social life if he spent so much time with Abby, who was so highly peripheral to the group's activities at this point in her life, but he did gradually become assimilated again.

In February 2003, when Nobu was just eighteen months old, Abby died and poor Nobu was once again in need of a mother. By this time his growth was severely stunted compared with his cohort. He had also lost half of his tail in an accident or a fight (which we did not see), and this made him seem even more pathetic. His diminuitive size, however, added to his charm, and as he emerged from his depression and became more social, Nobu became the darling of the group—the infant whom all females were most willing to indulge with rides, milk, and grooming. Even his agemates were willing to give him rides from time to time. By the time he was three years old, he was probably the most popular juvenile in the group, and despite his small size, he was able to hold his own in minor squabbles over foraging sites, winning out over even the alpha female's daughter.

In addition to nursing, the other potential advantage of capuchins' unusually long juvenile period and extensive exposure to alloparents is the opportunity for social learning. When they are just a couple of months old, infants start to show an interest in the foraging activities of others. Not content with watching casually from a distance, they run up to within just a few inches of a forager, sometimes even crawling onto him or her, and stare intently at the food and at the hand movements and the mouth of the forager.[14] They sniff and touch the food, sometimes even inserting their fingers into the mouth of the forager.

Foragers are generally tolerant of these intrusive observations, and sometimes they will even permit the infant to have a small taste of the food. Infants seem particularly excited when they have opportunities to watch adult males forage; they often squirm, threaten the food items, and have erect genitalia while observing them forage. It is still not entirely clear what they are learning or how much of an impact social learning has on their ability to acquire foraging skills; these topics are the subject of my current research (see Chapter 12).

After we explored these theories, capuchins' obsessive handling of small infants remained a mystery to us. Yes, there were obvious advantages to allonursing, but the handling itself did not seem to provide any advantage to the infant and probably entailed a small cost, since it disrupted the infant's sleep and nursing. Perhaps the infant handling is crucial to forming relationships between infants and alloparents, groupmates who will provide other services to them later. We were surprised to find, however, that the females who handled each infant the most during its first three months were not the ones who nursed the youngster most frequently when he was older.

Among the existing explanations for infant handling, the one best supported by Joe's analyses is the by-product hypothesis. The story of Tattle's baby Eldritch is one relevant piece of evidence. Unlike most infants, who arrive in clusters within a few months of each other, Eldritch was born when the next youngest infant in Abby's group was already eight months old (long past the peak "cute" stage), and the group's next infant was not born for more than a year. Eldritch thus received lots of handling—twice as much handling from the group's females as Tattle's previous infant, Fester, who had been part of a cluster of five infants born within a two-month period. It seemed as if attention to other females' infants formed a behavioral "pool" of approximately constant size (per female), and that Eldritch "absorbed" all the attention that Fester had had to share with four other babies. Eldritch's magnetic pull on the group's adult females offered further evidence against the schmoozing hypothesis, since Tattle was the group's lowest ranking female and usually quite peripheral.

The by-product hypothesis also explains why adult females are so eager to handle each other's babies that they "pay" for the privilege with grooming. Among white-faced capuchins and many other primate species, mothers of small infants receive far more grooming from females than they did in the months before they gave birth. Indeed, Peter Henzi and Louise Barrett found that among chacma baboons, female-female grooming and opportunities to handle infants acted as commodities in a "biological market": when infants were scarce, handlers groomed mothers far longer than when infants were abundant, suggesting that both parties were aware of the effects of supply and demand on the "price" of access to an infant.[15] If infant handling were beneficial to infants, we would expect, in contrast, for mothers to solicit it and to "pay" handlers for their services.

As Joe refined his data analyses and interpretations following our 1997 field season, he was reminded of a somewhat obscure idea our mentors Barbara Smuts and Richard Wrangham had found intriguing. In 1977 the maverick Israeli zoologist Amotz Zahavi published a short note in which he argued that certain kinds of social behavior function to "test" social bonds by imposing on the recipient.[16] Zahavi acknowledges his family, including his dog, as the inspiration for his development of the theory. Consider: why does your dog jump up and place his paws on your shoulders when you arrive home in the evening? Why not just gently place his head next to your knee? Perhaps your reaction to the more noxious greeting gives him an accurate idea of your current feelings toward him, and therefore what he can realistically expect from you for the rest of the evening. If you scowl and order him down, he knows that your relationship with him currently has low priority. But if you take hold of his paws and start dancing with him, he knows it's his lucky day. In contrast, your reaction to a more neutral greeting would tell him a lot less. If you doubt that being jumped on by a dog is inherently unpleasant, imagine being jumped on by a dog you don't know personally. Only your relationship with your own dog turns the experience into something tolerable or even pleasant. It's easy to think up other possible examples of bond-testing impositions, from the whining of children to the hazing rituals of fraternities. Zahavi points out that anyone will gladly accept a *benefit* during a social interaction, regardless of who confers it, but only someone who values his relationship with the other interactant will accept a *cost*.

Zahavi is best known for formulating, in 1975, "the Handicap Principle," which holds that costly, wasteful ornamental traits used in mate choice (of which the peacock's tail is the archetypal example) evolved precisely because of their detrimental effects on survival: they tell a prospective mate, in effect, "I've survived despite this handicap, so I must have excellent genes."[17] Zahavi presented all his ideas in words only, eschewing the formal mathematical models that evolutionary theorists rightly demand. A theory can sound seductively good but turn out to be unworkable once one has laid bare its assumptions and specified exactly how its hypothesized processes would affect the relative frequencies of alternative traits. After several attempts by other theorists to model the Handicap Principle failed, it and its originator were marginalized for several years. Then, in 1990, Alan Grafen produced a highly esoteric but rigorous mathematical account of how it could work, after all.[18] Today, under the label "costly signaling theory," the Handicap Principle is used to explain and predict many biological phenomena (though very few biologists accept Zahavi's claim that it explains virtually *all* animal, plant, and even intercellular communication). Zahavi's theory that social partners test their bonds by imposing on each other has the same paradoxical flavor as the Handicap Principle, but no one has yet tried to produce a formal model. Partly for this reason, almost no research had been based on the bond-testing theory before 1997, when Joe was working on the infant-handling data.

Perhaps capuchin infant handling functions to test bonds between females, just as "diddling" tests bonds between male baboons.[19] It must require considerable trust to allow another female, even a relative, to prod and poke your infant while it's on your back and therefore out of your view. And even if a handler means well, she may harm a baby accidentally or infect it with a disease. Female capuchins often support each other in aggressive coalitions, so maybe they communicate about their commitment to defending each other through their reactions to others' attempts to handle their infants. This hypothesis superficially resembles the schmoozing hypothesis, with a subtle yet important difference. The schmoozing hypothesis, which holds that infant handling *strengthens* female-female bonds, requires that handling *benefits* infants—something for which we found no evidence. But Zahavi's theory implies that infant handling *tests* female-female bonds by imposing a small *cost* on infants— a cost that a mother is willing to bear only if she is currently strongly committed to her relationship with the handler.

The bond-testing hypothesis leads to three testable predictions.[11] First, females will most frequently handle the infants of their most frequent coalition partners. This turns out to be true, and there was no tendency for females to handle the infants of females *against* whom they formed coalitions more frequently (another blow to the harassment hypothesis). Mothers sometimes call a halt to infant- handling episodes by leaving the scene or by turning their back so as to take their babies out of reach of their admirers, and the second prediction of the bond-testing hypothesis is that mothers will tolerate longer episodes of infant handling when the handler is a female that they already have a close bond with, as measured by the two females' rate of grooming *before* the infant's birth (remember that moms of new infants receive extra grooming just because they have a baby). Our data provided only limited support for this prediction; different statistical techniques gave different results. Third, when the mother and handler were frequent grooming partners before the infant's birth, a higher proportion of handling episodes will involve actual touching of the infant, as distinct from mere close inspections. The reasoning here is that only a mother with a strong bond with the handler will permit the handler to put her infant at greater risk of harm by actually touching it. In the data there was a strong trend in the predicted direction, but it failed to reach conventional levels of statistical significance.

On the basis of these results, we could conclude only that bond testing might be a function of infant handling in capuchins, but that the byproduct hypothesis is at least as well supported. Nevertheless, we became fascinated by the potential of Zahavi's ideas for explaining puzzling forms of social interaction—of which our monkeys provided plenty (see Chapter 11).Another enigmatic capuchin signal that may reflect Zahavi's ideas is the gargle vocalization, which is produced primarily by infants and their mothers.[20]

The gargle is a loud, obnoxious-sounding, raspy vocalization produced by females and infants when they are highly aroused and in close range of adult males. It is one of the most puzzling vocalizations in the capuchin's vocal repertoire and has no analogue in other primate vocal repertoires, to the best of my knowledge. Adult females gargle only to the alpha male, and they do so primarily when they have a newborn infant. The gargle is the first or second vocalization to enter the infant's repertoire, beginning at about two weeks of age, and during the first three months of an infant's life the mother and infant will perform gargle duets while obsessively following the alpha male. These duets are striking because the mother gives the impression that she is actively teaching the infant the appropriate context for gargling. This is the only behavioral context in which I have ever felt that a capuchin mother was attempting to teach her infant anything. The mother-infant pair is always enthusiastic in its gargle duets, but the alpha male does not seem particularly pleased to receive these attentions: either he ignores them completely, or he swats and nips at them.

Once an infant has firmly established a pattern of gargling whenever it nears the alpha male, the mother's gargle rate declines and she no longer gargles regularly with her infant. Infants continue to gargle at high rates once they are moving about on their own, and they gargle to all adult males, not just to the alpha. Infant gargle rates peak at four to six months of age and then rapidly decline with age; grown males do not gargle at all.

We do not yet understand the function of gargling. It clearly does not work well as an appeasement signal, since it often annoys the male it is directed toward. And it seems unlikely to serve as a signal of allegiance, since (1) alpha males do not seem at all perturbed to hear group members gargling to other males, either in playback experiments or in naturalistic contexts, and (2) there is no plausible reason why males should care about the willingness of helpless infants to support them. My best guess is that gargling tests the calling monkeys' bond with the adult male recipient. In other words, this irritating stimulus may be produced by young infants and their mothers to test the goodwill of the male, so they will know how they stand with him. The vocalization is produced primarily when the infants are still at an age when they could be vulnerable to infanticide, and so it is critical for them to know which males find them irritating and which males enjoy their company.

In what way do capuchin mothers help their offspring? In many species, one or both parents play a protective role, shielding infants from aggression by peers, adults, and would-be predators. Capuchin mothers are less protective than mothers of many primate species, allowing their young to socialize with practically anyone in their group and to investigate predators at shockingly close range. For example, they will permit their infants to toddle up to within a meter of a dangerous snake and—although they will alarm call like the rest of the group—typically will make no effort to retrieve the infant or to modify its behavior. The same is true for mammalian terrestrial predators: mothers act as if their infants are perfectly capable of fending for themselves against canids and felids. They do, however, react quickly if a very young infant falls out of a tree and lands on the ground near a human observer: in this case they will utter a quiet alarm call and hastily run to the ground to retrieve the infant. When very large raptors fly overhead, females will rush to their infants and invite them to get on their back. They will also sometimes offer rides to other females' infants in such circumstances.

The situation that elicits the most panic in a capuchin mother is the sighting of an unfamiliar male. When intergroup encounters begin, the females make a beeline for their infants, insist that they jump on their back, and then bolt for safety. Clearly they view unfamiliar capuchin males as dangerous to their babies. Occasionally even relatively familiar capuchin males are treated with suspicion as well: for example, Diablita diligently kept her infant Frodo away from Tranquilo, and Cupie kept her infant Chaos away from Pablo. These anomalous situations are highly noteworthy, given capuchin mothers' generally casual attitude about letting others handle their offspring. It seems that Diablita and Cupie had tagged those two males as being particularly likely to commit infanticide. Both females treated the same males differently when they had infants that were fathered by them, which suggests that the females have a fine-tuned means of calculating the risk of infanticide.

In many primate species, such as most macaques and baboons, mothers help their daughters acquire a dominance rank just below their own.[21] So far as we can tell, in capuchins neither parent aids its offspring in attaining high rank. We are still waiting for the subjects of our developmental study to reach adolescence, when dominance rank stabilizes to some extent. However, analysis of earlier data (albeit from a sample size of only seven mother-offspring dyads) indicates that females actually intervene against their offspring more often than they intervene in their favor. In those cases in which mothers supported their young offspring, the offspring were fighting males. When offspring were fighting females, the mothers supported them against lower-ranking infants but regularly intervened against their own offspring when they were fighting against females who occupied a higher rank than their offspring. This evidence indicated that they were "teaching" their offspring their place in the hierarchy, rather than helping them to ascend to a higher level in that hierarchy.

Taking a human perspective, we might expect capuchin mothers to actively teach their infants what to eat and what not to eat. However, as in many other species of primates, capuchin mothers let their infants eat anything they want, and they never try to prevent them from sampling inappropriate foods.[22,23] Every time I see an infant handling cashew fruits for the first time, I hold my breath, hoping that the youngster will not try to eat the highly poisonous seed. The forest is full of toxic items, and it really seems miraculous that infants and juveniles do not poison themselves more often. Although they gain plenty of information by watching others forage, the infants are entirely responsible for deciding which individuals to watch, and when. No one attempts to modify their foraging behavior when they make mistakes.

Whereas capuchins do not try to modify others' foraging behavior, they very eagerly attempt to modify one another's social behavior. Capuchins are highly interactive, opinionated, and meddlesome creatures. Most aggressive, play, and sexual behavior inspires comment from other groupmates, many of whom often seem truly outraged. Although we are only now beginning to examine the role of social influence on the development of communicative behaviors in capuchins, it seems highly probable that individuals' attitudes, modes of communication, and social tactics are in large part the product of their social environment. Monkeys probably learn about their own rank, and about appropriate modes of interaction with monkeys of different ranks, both by observing others' interactions and by receiving punishments when they make social blunders themselves.

# CHAPTER 11

# Guapo: Innovation and Tradition in the Creation of Bond-Testing Rituals

It is noon on a scorching-hot April day in 1992. Joe and I are with Abby's group, doing a follow of Abby herself. As usual, she is resting with her best friend, Squint. This pair of monkeys makes a formidable team: Abby is the alpha female of the group, and Squint is constantly at her side, meddling in the affairs of all the other monkeys and recruiting Abby's aid in keeping "order." They are particularly hard on males. Whenever a male so much as looks at Squint the wrong way, she screams in outrage and solicits Abby's help in evicting him from the tree. But at the moment Squint is not interested in causing trouble for anyone. They are seated on a branch some distance from the other monkeys, and Squint is peaceably grooming Abby. Then she pauses and cups her hand over Abby's nose and mouth. They sit this way for a while, gently swaying, facing each other but not making eye contact. After forty-five seconds or so, Abby rouses herself and grooms Squint for a few seconds. Squint seizes Abby's hand and places it over her own nose, leaning back and shutting her eyes halfway. They remain in this pose for about three minutes, and then Squint puts a hand over Abby's nose. They look like they are in a trance.

After watching this for about ten minutes, Joe and I start to giggle. As often as we have seen this behavior, it still seems strange when the monkeys, who are usually engaged in frantic, rapid motion, stop and focus for so long on anything, especially something this silly-looking. "Vulcan mind meld," says Joe.

"But seriously, what can it be for?" I ask. This question was to lead me down a long sequence of rejected hypotheses and result in a major and unexpected shift in my research, though I had no inkling of this at the time.

Initially, the most likely seeming explanation for this cupping of hands over another's nose—what we term hand sniffing—was that it was some sort of olfactory communication. But what could the monkeys be communicating? Were they sharing information about what they had just eaten? If so, one would expect juveniles to sniff adults' hands more often than adults sniff adults' hands, because juveniles need to learn the most, but the data didn't fit that assumption. Only adult female dyads had been observed sniffing hands.

Perhaps the females were synchronizing their ovarian cycles, as discovered by Martha McClintock in her studies of rats and of women living in dormitories.[1,2] By cupping their hands over each other's noses, and—in some cases—sticking their fingers way up the nose, female capuchins might be transmitting pheromones that influence their reproductive cycles. We had no idea whether the monkeys' cycles were synchronized or not, since there is no external sign of estrus in this species. There was a fair amount of birth synchrony, but that could easily be explained by other factors, such as seasonal differences in nutrition.

Why would it be advantageous for females to have their cycles synchronized? Perhaps they could have more freedom of mate choice if they were in estrus at the same time. In this situation, the alpha male would not be able to guard all of them at once, so females would have more freedom to choose their own mates. This sounded like a reasonable adaptive explanation, but we didn't have the information to test it yet, so we filed it away for future pondering. "So if they are synchronizing their cycles, with whom are they synchronizing them?" asked Joe.

"That would be Nanny," I laughed. Most pairs of hand sniffers engage in mutual hand sniffing, with each monkey having the partner's hand on or inside her nose, either simultaneously or in turns. But one female,

Nanny, stoutly refused to sniff anyone else's hand. She almost always sniffed her own hand while sticking the fingers of her other hand up a friend's nose. The following sequence, recorded in March 1993, describes Nanny hand sniffing with her most frequent sniffing partner, Squint (though Squint did not always have this much trouble adjusting the position of Nanny's finger).

> Squint and Nanny groom for a while, and then Squint grabs Nanny's hand and inserts a finger up her own nose. Nanny cups her own right hand over her own nose. Then Nanny removes the hand from her own nose and grooms Squint with just one hand, because Squint continues intently sniffing Nanny's left hand. Suddenly, Squint pulls Nanny's finger out of her nose and frantically reaches up her nostril, trying to pull something out of it—it appears that she may have jammed Nanny's finger too far up. Nanny gently rests her left hand over Squint's nose and sticks her right index finger up her own nostril. Squint removes Nanny's hand while she once again attempts to set matters right within her nostril, and then reinserts Nanny's finger far up her own nose. Once again, she removes Nanny's hand, tries to remove something from her nose, and stubbornly reinserts Nanny's finger. Squint sneezes, blasting Nanny's finger out of her nose, and promptly sticks it back in again. Finally she seems to have solved the problem, and they settle into a long, relaxed sniffing session, with Squint sniffing Nanny's finger while Nanny sniffs her own hand.

Observations such as this one highlighted to us how hand sniffing can be uncomfortable. Capuchins' fingers are pretty large in comparison with their nostrils—putting a finger up a nostril is a tight squeeze, and makes breathing difficult. Also, the monkeys have long, sharp, clawlike fingernails. Why on earth, we wondered, would they want someone else's claw lodged inside their delicate nasal passage, when one startled movement by either partner could give them a nosebleed? Yet the monkeys seemed to crave this behavior, almost as if they were addicted to it.

Hand sniffing brought to mind the work of Nancy Nicholson, who hypothesized that humans like to kiss their partners so much because they become addicted to their sebum (an oily compound secreted from glands in the skin).[3] Could this sort of process explain the monkeys' fascination with sticking friends' fingers up their delicate nostrils? Perhaps, but it did seem that it would be more effective to pair two delicate skin surfaces more effective for transmitting sebum, rather than a dirty, calloused fingertip and a nose's tender interior.

Intrigued by this behavior, I asked my colleague Linda Fedigan, of the University of Alberta, whether she saw it in her monkeys. Linda had been the first person to study social behavior in white-faced capuchins, and her study site, Santa Rosa, was only about fifty kilometers from Lomas Barbudal. Linda wrote back and told me that she had seen it in one of her groups at Santa Rosa—but in that population, it was almost exclusively adult males who did it. So much for our reproductive-synchrony idea. It was quite puzzling that hand sniffing was an almost exclusively male behavior at one site and an almost exclusively female behavior at another.

When I was writing up my early research data, I began mulling over the problem further. Were there differences in the social organization of the two study groups that could help explain this behavioral difference? In Abby's group, the females were quite tightly bonded with each other in comparison with the males. Using grooming as an indicator of the quality of social bonds, the difference in the quality of female-female versus male-male bonds was not nearly so great in Santa Rosa's CP group, apparently because the males were more tightly bonded there than at Lomas. Perhaps hand sniffing played a role in creating social bonds? I looked at the patterning of intragroup variation in Abby's group and found that the pairs that groomed the most often also hand sniffed the most often. This was consistent with the hypothesis that hand sniffing cements social bonds, but it did not help to clarify its role in the bonding process.

After completing my dissertation work, I went to the University of Alberta on a postdoctoral fellowship to collaborate with Linda Fedigan and her new student Kathy Jack, who was soon to begin work on the Santa Rosa white-faced capuchins. I had already met some of Linda's former students, including Lisa Rose and Katie MacKinnon, while visiting other sites in Costa Rica. Linda's students were all wonderful people, and we had a great time swapping monkey stories. We were keen to collaborate on something but had not yet settled on a topic. In the meantime, we compared our methodologies and our definitions of behaviors, and made sure we were using the same terms for particular behaviors. We were especially interested in documenting the contexts of rare behaviors, for which one site's data might not yield an adequate sample size for analysis.

Before I left Alberta, Linda and I set up an e-mail network to foster discussion among capuchin researchers, including researchers of other capuchin species as well as those working with white-faced capuchins. To start the discussion, I sent around a message describing some of the more strange behaviors I had seen and inquiring whether others had seen them too. I was beginning to suspect by this point that some behaviors were "cultural"—that is, that they were not a standard part of the species-typical behavioral repertoire but were innovations that had been passed on via social learning in particular groups. I asked other researchers to nominate odd behaviors they had seen as well. This sparked a flurry of excited messages as researchers from several sites contributed their observations, and some interesting patterns began to emerge.

There were some behaviors that appeared to be "capuchin universals," despite their peculiar form and lack of obvious utility. This list included, for example, the capuchin habit of rubbing the nose vigorously on rough bark and then sneezing before lying down. It also included the frenzied rubbing of various noxious substances into the fur as a social activity (and a possible insect repellant). Various forms of play also appeared at most or all sites. In one common game, capuchins hang by their tail and swat at each other. In a common variant, one monkey hangs by its tail and holds another monkey's tail in his feet, wrestling him with his arms until he tires of the game and lets go of the partner's tail, dropping him unceremoniously to the ground from a great height. Various displays were also widely distributed, such as the "tail branch drag," in which a male seizes a leafy branch with his tail and charges through the dry leaf litter with it, creating a terrible racket.

There were many behaviors, however, that were particular to just one or two social groups. Some were clearly innovations created and practiced by a single individual or pair of individuals. Others were more widely distributed within a social group but absent in most other groups and sites. We decided to focus on the latter set of behaviors and to conduct a thorough investigation of these behaviors for a joint conference talk. As our discussion progressed, Doree Fragaszy, a psychologist working on social learning in brown capuchins, convinced us that the topic was exciting enough to warrant an entire conference of its own, not just a single conference talk at the annual meeting of the American Society of Primatologists. Doree and I wrote a grant proposal to fund a conference on traditions in capuchins that was held at the University of Georgia—Athens in 2000, and the papers presented there evolved into an edited volume.[4] Since most of us were quite new to the topic of social learning and traditions, we invited experts working on other taxa (apes, macaques, dolphins, rats, and birds) to attend the conference as well, to give us a wider perspective.

From its beginnings as a small project somewhat peripheral to my primary research, my study of social learning in capuchins had grown to the point that I needed to work with collaborators to treat the topic properly. I recruited Melissa Panger and Lisa Rose to help with the data analysis. Melissa, who had conducted detailed work on foraging techniques with the white-faced capuchins of Palo Verde, Costa Rica, was to coordinate the cross-site integration of analyses of behavioral variation in food-processing techniques,[5] and Lisa Rose, who had written about predation on vertebrates, was allocated the job of coordinating the analysis of intersite differences in interactions between capuchins and other species.[6] This allowed me to devote more time to my favorite topic: the quirky social conventions like hand sniffing.[7]

We decided to include data from four different study sites: Lomas Barbudal, Santa Rosa, Palo Verde, and Curú. All are in northwestern Costa Rica, and therefore they are similar ecologically (though Curú is on the coast and has some species not present at the other three sites). Furthermore, because they are fairly close together there was good reason to believe that the four capuchin populations were similar genetically. Until about fifty years ago, the sites would have been connected by forest corridors. Therefore any behavioral differences between sites were quite likely to be due to "cultural" differences—in other words, social learning was likely to be the mechanism responsible for maintaining differences between groups.

This project, like all collaborative endeavors, took several years to complete. For the social-conventions study alone, we painstakingly waded through more than 19,000 hours of behavioral data (really more like 25,000 hours, since there were multiple observers collecting data on the same groups during many of the time periods). From these observations, we collected all reports of hand sniffing and other quirky social interactions that were not part of our standard ethograms. We designated a behavior as a tradition only if it was quite common in some social groups or sites yet completely absent in some other groups that had been thoroughly studied.

Our data on hand sniffing in thirteen social groups, collected over a period of up to eleven years, painted a picture that was even more complex than we had known. Hand sniffing was present in some form in five of the thirteen groups, and these five groups lived at three different study sites. At the sites that had more than one well-studied group, there were some groups in which hand sniffing was never observed.

Hand sniffing was not a permanent part of the repertoire in any group. For example, hand sniffing was very popular among females in Abby's group for a period of seven years, but when two of the most avid hand sniffers (Squint and Wiggy) died, the hand sniffing tradition died out as well. Nanny and Vandal would still occasionally sniff their own hands in the midst of grooming, as if "reminiscing" about the old days, but social hand sniffing disappeared from the repertoire. Santa Rosa's CP group had an odd pattern of hand sniffing as well. In 1986 the males often sniffed one another's hands and occasionally sniffed females' hands as well, but then the tradition vanished. It reappeared in 1996 when it was reintroduced by an adult female, who mainly hand sniffed with the alpha male. Apparently these hand-sniffing traditions were quite fragile and could easily be disrupted by the disappearance of key members of hand-sniffing cliques.

The really puzzling thing about the geographic distribution of hand sniffing was that it occurred at three different sites yet was absent from well-studied groups that had regular contact with groups containing hand-sniffing cliques. Since migration between sites was not possible (except, perhaps, between Lomas Barbudal and Palo Verde, which still had a tiny corridor connecting the two areas), it seemed that hand sniffing must have been invented independently several times. At first this sounded implausible to me, but on second thought the idea did not seem so strange. After all, new inventions must be created from "building blocks" consisting of smaller behavioral units. Capuchins are extractive foragers *par excellence*, and they find it perfectly natural to probe with their fingers any small holes they encounter, looking for insects and grubs to eat. Their attention is thus likely to be drawn to the nostrils and mouths of their companions—and the desire to pick a friend's nose is not so hard to explain. The difficult bit is understanding why the friend does not respond more negatively to having a sharp fingernail in his or her nose.

Hand sniffing was not the only candidate for the "tradition" designation. There were other strange rituals as well. One that cropped up in multiple places was the habit of sucking on body parts, such as tails, ears, and fingers. This behavior was seen primarily in males who were either migrating together or on the verge of migrating. One pair of males at Santa Rosa that migrated together between three different groups engaged in frequent hand sniffing and finger sucking interludes, in which the dominant male was the sucker in all interactions.

Guapo from Abby's group was a frequent sucker when he was a subordinate male, and he did this in two different contexts. From 1990 to 1996, Guapo sucked other monkeys' tails as a way of inviting grooming. This signal was initially misunderstood by his partners, who would frantically try to pull their tail out of his mouth, but he gradually "trained" them to understand the signal by refusing to let go until he had received adequate grooming. Later, when a new immigrant male (Hongo) arrived and he and Guapo were the only subordinate males living with a particularly despotic alpha male (Ichabod, who was subsequently killed by the members of his own group), Guapo and Hongo sometimes sneaked off to the edge of the group together and mutually sucked each other's tails for extended periods of time.

Sucking was most rampant in Rambo's group in 1997, when approximately one-third of the group was seen to be involved in the sucking of ears or tails. In this group, unlike the observations from most other groups, sucking was quite often mutual, and could last for an hour at a time. However, the vast majority of these incidents involved one particular adult male, Q, and his most frequent sucking partner was Rambo, another male with whom he frequently made long excursions outside the group, as if they were contemplating comigration. When the primary sucker disappeared from the group, the sucking tradition essentially vanished from Rambo's group as well.

In Abby's group in 2003, two immigrant males (Took and Bentley) spent astounding amounts of time tail sucking, with Bentley always sucking Took's tail (or ear) for extended periods. We do not know their history, but it is possible that they had migrated from the same group, even though they arrived in Abby's group at different times; they look strikingly similar. Following is a typical tail-sucking incident involving this pair:

**May 20, 2003.** It is three days since Took and three other males immigrated into Abby's group. It is sheer chaos, with all the males frantically monitoring one another and threatening one another; no one is sure who is on whose side. Took and Bentley have just formed a coalition against Weasley, the alpha male who was wounded during the immigration three days ago. Bentley wheezes at Weasley, who grunts in response from a distance, but then Bentley lies down beside Took and puts Took's tail in his mouth. The pair of them seem completely at ease with each other. Took begins to doze, and Bentley glares at me, still perhaps riled up from the fight with Weasley. Bentley removes Took's tail from his mouth, handles it thoughtfully, threatens me, and reinserts the tail in his mouth. Took rests, his head turned away from Bentley, apparently oblivious to the fact that Bentley just threatened me. Bentley tugs on Took's tail a bit, as if trying to rouse him to join in a coalition against me, but Took is too relaxed to respond. Slowly Bentley calms down, still sucking on Took's tail, and closes his eyes. They lie like this for over twenty minutes, with Bentley occasionally rousing himself to stuff the tail even further into his mouth. Took has an unusually fat, fuzzy tail, and it appears to be a big mouthful.

Five minutes after the sucking incident ends, adult female Opie runs over to them, and she and Bentley threaten me. Then Opie begins to direct vocal threats at Took, and she headflags to Bentley for aid. Normally an adult male wouldn't hesitate for a second to aid an adult female against one of his reproductive competitors, but Bentley declines to join in threatening the male whose tail he has just been sucking.

The most bizarre ritual by far was one that we discovered in Pelon group in 2003: eyeball poking, which typically occurs in female-female and male-female pairs. In this ritual, one monkey pokes another in the eye, inserting the finger into the eye socket up to the first knuckle. The monkey being poked typically holds onto the partner's hand, and may even initiate the eyeball poking by inserting the claw-like fingernail of

Adult females Rumor and Tamora interlock their hands while inserting their fingers up each other's nostrils, following an eyeball-poking session. Photo: Susan Perry.

the partner inside the eyelid and gently squeezing it between the eyelid and the eyeball. Monkeys in Pelon group sometimes poke themselves in the eye while hand sniffing, almost as if they are trying to get up the nerve to receive poking from the partner. The exact form of the eyeball-poking ritual varies from dyad to dyad: some monkeys stick their fingers in their partner's nose or mouth while receiving eyeball poking, while others poke themselves in the eye.

Eyeball poking is in some ways the most astonishing of the capuchin rituals we have observed, because the risk to the eye is so great. Capuchins' fingernails are sharp and dirty, and are covered with food remains as well as a partner's mucous and saliva. The monkey being poked is subjecting itself to the risks of infection and impaired vision, should the cornea be scratched. It is obvious that the one being poked is also undergoing some physical stress, since he bats his eyelids constantly during eye poking and also twitches the corners of his mouth when the finger moves in the eye socket. Yet the monkeys remain enthusiastic about being poked, demanding reinsertion of the finger if it pops out of the socket. These eye-poking sessions can last for up to an hour.

Perhaps the most interesting of the social traditions performed by the monkeys was the class of behaviors we termed "games." Guapo was exceptionally boisterous, creative, and playful when he was a young subordinate adult male in Abby's group, and he invented several games. At first I did not see anything unusual about this—capuchin monkeys are playful and intelligent animals, I thought, so why shouldn't they make up new games? But when I talked to primatologists who studied other monkey species, I always encountered surprised interest when I described these rituals; they said that their study subjects never deviated from the rough-and-tumble play (mainly biting, chasing, and wrestling) typical of most primate species. So I went back through my data and pulled out every description of a game I could find, sorting them into types and noting who played with whom and which role each player performed.

The most common game was the "finger-in-mouth" game. A description of a typical episode follows:

January 25, 1992. Guapo and Kola (a juvenile male who is approximately three years old) are on the edge of the group. Kola grooms Guapo with one hand, while one finger of his other hand is lodged in Guapo's mouth. When gentle tugs fail to free his hand, Kola braces his foot against Guapo's chin and tugs harder, to no avail. Using his free hand, he parts Guapo's lips, inspects his teeth, and tries to pry his mouth open. Once again he attempts to use his foot for leverage, this time bracing it against Guapo's shoulder while he pulls on his stuck hand. Kola does not seem at all distressed—rather, he acts as if he is attempting to solve an intriguing puzzle or foraging task. After a prolonged effort, he gets his fingers out of Guapo's mouth. Rather than leaving, he resumes grooming Guapo, using both hands. He grabs Guapo's hand and grooms it for awhile, and then Guapo once again grasps Kola's finger and clamps it between his teeth, beginning the game anew.

Guapo plays the finger-in-mouth game with a juvenile in 1992. Photo: Susan Perry.

The variation of Guapo's game that I found most amusing was the "hair game."

November 29, 1992. Hobbes, a three-year-old juvenile male, is grooming Guapo on the periphery of the group. Guapo leans forward lazily and bites a big wad of hair out of Hobbes' cheek. It looks like it must be painful, but Hobbes barely flinches as the roots are pulled out. He immediately gets to work trying to retrieve the hair from Guapo's mouth, sliding his fingers between his lips and tugging at the protruding hairs, which are clamped between Guapo's teeth. Hobbes gargles at Guapo and grooms him for a bit before resuming his efforts. Finally he succeeds in opening Guapo's mouth just enough to pull most of the hairs back, and he clamps them between his own teeth. Guapo tries to retrieve them, but fails. Instead he bites another large tuft of hair out of Hobbes. Hobbes pries at Guapo's mouth again, alternating between grooming and tugging on the hairs. Finally Hobbes bites a big tuft of hair out of Guapo's face.

This game can continue for up to thirty minutes, and I have on occasion seen as many as twelve large tufts of hair removed from a monkey during one game. Sometimes the monkeys will alternate between playing the finger-in-mouth game and the hair game.

The other major variant among Guapo's games was the "toy game." This game was played in much the same manner as the others, except that another object took the place of the finger or hair tuft as the thing that was being extracted from the mouth. The object was always an inedible object, such as a leaf, a twig, a bit of bark, or a green fruit. The toy was never eaten; after passing it back and forth, it was always discarded in the end.

> **February 11, 1993.** Guapo is sitting with juvenile male Hobbes. Guapo picks some leaves and puts them in his mouth so that they protrude slightly. Hobbes tries to pry Guapo's mouth open with his hands and mouth, and he finally succeeds in transferring the leaves into his own mouth. Then they switch roles a couple of times, each trying to remove the leaves from the other's mouth, until they have been dropped. Guapo picks another leaf and a piece of bark and puts them in his mouth. They are rapidly retrieved by Hobbes. Guapo snatches them back from Hobbes's hand and resinserts them in his own mouth so that they protrude enticingly from between his teeth. Guapo holds onto Hobbes's hand as Hobbes tugs on the bark and leaf. Guapo pulls gently on one of Hobbes's hands while Hobbes uses the other to pull the "toys" out of Guapo's mouth . . . eventually they tire of this game and let the bark and leaf fall to the ground. Unlike a food-theft incident, in which the monkeys would be fast-moving, aggressive, and screaming in outrage, this interaction has a relaxed nature, with both monkeys calmly, quietly, and enthusiastically focused on the task of opening the mouth and retrieving the leaves.

In 1991 and 1992 Guapo played these games almost exclusively with the juvenile males who were his most frequent playmates: Kola, Quizzler, and Hobbes. As Guapo's special infant friends Rico and Lizano matured, they also became game players. Females played these games only on rare occasions; the games were typically a male-male behavior, just as hand sniffing was a female-female affair in Abby's group. As Guapo's little friends matured and new cohorts of infants matured to game-playing age, Guapo's juvenile playmates began to play the games with the younger males.

In 1999, when Guapo became alpha male of Abby's group, he stopped playing games, but the tradition persisted among the younger males of the group. Kola, Hobbes, and Quizzler, the trio of males with whom Guapo had played games most frequently, emigrated in 1996, and Kola and Quizzler were part of the same all-male group for awhile. When Kola returned to Abby's group in late 1998, he once again resumed the games with the younger males Rico, Cimo, and Dos. After Guapo stopped playing, all games included those four males. In 2001 the four males emigrated, forming an all-male group together. We spent a little time following this all-male group and were able to confirm that they were still playing the hair game. Thus, the tradition continued elsewhere, even after it was extinguished in Abby's group. Both Kola and Rico died in 2002, but Cimo and Dos migrated together into a new group. While the females in that group are still too unhabituated for us to be able to collect behavioral data, we are anxious to know whether Cimo and Dos transmitted Guapo's game-playing traditions to their new group.

I was intrigued to hear that the toy game was independently invented in another group of capuchins living in Curú. Mary Baker had a very nice data set describing the toy game, which was played by approximately half of the members of her study group. The game-playing dyads always included at least one adult, who was playing either with a juvenile (sex often unknown) or with an adult of the same sex. While the distribution of the behavior across the social network was somewhat different at Curú than at Lomas (the games at Lomas were played disproportionately more often by adult male–juvenile male and juvenile male–juvenile male pairs), the form of the game was strikingly similar.

As with the hand sniffing tradition, it is perhaps not so surprising that the toy game was invented in two locations, given that all capuchins have evolved as extractive foragers, keen to work hard at removing food items from whatever crevices and matrices in which they are lodged. The hair game was also apparently invented a second time by Napoleon, an immigrant in Flakes group, in January 2006. Napoleon plays this game with Jade and Toulouse, the juvenile offspring of females who were in Abby's group when Guapo was playing the hair game. However, so far as we know neither juvenile male had had any opportunity to witness the hair game before Napoleon introduced it to them.

Were these games and other social conventions just pure fun? Or did they serve to communicate some important message about the participants' relationships to one another or to other monkeys that might be watching? A feature of both hand sniffing and some of the games that really struck me was that they looked like they would be uncomfortable, potentially risky, or even downright painful, yet the monkeys seemed to really enjoy the interactions. I was reminded once again of Zahavi's idea that some aversive signals can be used to test the goodwill of comrades.[8] These signals would actually be perceived as pleasurable if imposed by a good friend, and disgusting or otherwise aversive if imposed by someone that the recipient did not like. (See Chapter 10 for a more detailed description of this idea.)

Barbara Smuts and John Watanabe had proposed that male olive baboons test their relationships with alliance partners by handling each other's testicles (see Chapter 5).[9] These males' relationships are tense— for example, they never groom each other—but certain pairs often team up to take sexually receptive females away from individual males that outrank them both. This is, of course, a risky endeavor: a male abandoned by his coalition partner in the middle of a fight against a higher-ranking opponent is in serious trouble. Smuts and Watanabe argued that the trust necessitated by intimate physical greetings—in which the two males literally hold each other's future reproductive success in the palms of their hands—enables coalition partners to trust each other during fights. And indeed the pairs of males that handled each other's testes most frequently were also the most successful at teaming up to take females away from higher-ranking males. Perhaps a similar logic could be applied to behaviors such as hand sniffing, eyeball poking, sucking, and the hair game.

The next question, however, was: Why should these actions be culturally transmitted, highly variable behaviors, rather than part of the species-typical repertoire? Wouldn't the behaviors be easier to interpret if all members of the species engaged in them and understood them? Capuchins clearly have many species-typical vocalizations and gestures in their regular communicative repertoire, for negotiating various aspects of their social relationships: wheezes, grunts, gargles, and non-conceptive sex, for example. I started thinking about what characterized the users of the more culturally variable signals, as opposed to the species-typical signals.

Any dyad might wheeze or grunt together, but we had seen that these vocalizations were most common in dyads whose social relationship was under stress, either because the relationship was just forming or because some circumstance, such as a recent fight or the estrous status of the female, was producing tension. Gargles were produced most often by young infants and mothers of infants who were young enough to become infanticide victims. Nonconceptive sexual interactions were most commonly exhibited by male-male dyads—generally when the two males knew each other rather well but also were undergoing stress (for example, because of rank reversals or frequent interruptions of their affiliation by higher-ranking monkeys or new immigrants). In contrast, traditional social conventions, such as games and hand sniffing, were exhibited most often by dyads that already had a firmly established, fairly relaxed social relationship. This made sense—after all, it would be dangerous to walk up to someone you had just met and poke your fingernail in his eye socket or stick a finger between his teeth; he might misinterpret the signal and bite your finger off. It explained why culturally variable signals were not used by dyads that did not already have a somewhat solid relationship, but it did not explain why more stereotyped signals would not work as well.

I started thinking harder about signal design. Was there any aspect of the "design" of culturally variable signals that would make them a richer source of information than stereotyped ones? How could a variable signal be more "honest" than a stereotyped one? I could find no literature in biology to give me any insight—there was surprisingly little written on the evolution of signaling about quality of social relationships. So I turned to the sociology literature—in particular, to a series of fascinating papers by Randall Collins on what he termed "interaction rituals."[10,11] Because Collins studied humans, most of the interaction rituals he was thinking about were conversations, but any type of social interaction, whether it was gestural or vocal, would qualify. Collins argued that individuals living in complex societies have trouble gauging their precise status in a social network, because the relative strengths of dyadic social bonds are constantly shifting, and many group members' social interactions happen out of any particular individual's view. He suggested that people assess the status of their relationships, and their place within a wider social network, by gauging their own emotional state and that of their partners during interaction rituals.

The semantic content of these rituals is relatively unimportant—it can consist of useful information, complaints about the weather, wild speculation about who will win the next football game, outright lies, or sheer nonsense. The most important aspect of an interaction ritual is typically the emotional response of the other participant. Does she maintain appropriate eye contact, or does she gaze distractedly around the room while you are talking? Does her voice sound enthusiastic or bored? Does she maintain appropriate conversational pauses, butt in on top of you in a rude way, or fail to respond in a timely manner to your comments? Does she make or laugh at in-jokes that link this conversation to past ones you have had, thereby reaffirming the bond, or does she fail to pick up on your attempts to joke about things you have experienced together? All of these signals and many more can provide useful information about how the partner feels about your friendship.

The logic of interaction rituals made a lot of sense to me with regard to capuchin behavior. A signal that was innovative, rather than species-typical and stereotyped, would have to be learned by a particular dyad, and therefore would require a greater commitment of energy and time. As I was to discover later when studying the development of social conventions, it often took quite a long time for pairs to work out the details of a satisfactory convention in which each party was happy with his or her particular role. Signals such as the games, which involved multiple roles and turn taking, required even more focus and coordination than simpler signals.

A feature of all the traditional capuchin social conventions that really stood out was the degree of focus they elicited. This focus did not take the form of eye-to-eye contact, which in capuchins is exhibited primarily in moments of aggression and during sex, rather than in affiliative interactions. What was remarkable was how these normally frenetic, active monkeys would slow down and focus on each other without getting distracted during hand sniffing or game playing. And because the more variable and complex interaction rituals required greater focus, it seemed that they would be richer signals of commitment.

Part of a ritual's signaling value comes from the fact that the interaction is unique or virtually unique to a particular dyad—they have worked out a system of rules together. For this reason, I did not expect these traditions to last forever. The precise form of the games, hand sniffing, and other rituals would be expected to change subtly as new practitioners were added to the network of individuals sharing the practice and the new members added their own touches while working out the details with their partners. Social conventions would be expected to fade out as the composition of social networks changed. However, new ones would take their place, and although the precise motor details of these interaction rituals would perhaps be different, they too would be characterized by elements of risk or discomfort (so effective in eliciting variable emotional responses, depending on the nature of the relationship), and by complex or odd behavioral elements that would require a great deal of focus on the part of the participants.

All of these speculations about capuchin traditions and their functional origins made me think more about the evolution of human culture. Most anthropological research on human cultural origins has focused on technology and foraging efficiency (those aspects of culture that are most tangible and readily preserved in the archaeological record) rather than on features that are relevant to defining social relationships (such as ethnic and class markers, greetings, and social norms defining proper interactions with various members of society). This is also true of the "cultural primatology" literature. Traditions had already been reported in some other primate species, particularly macaques[12] and common chimpanzees,[13] but the vast majority of the potentially cultural traits that were described involved foraging innovations, such as food-processing techniques or food choices. Exceptions include a few communicative gestures in chimpanzees that are used in different contexts at different sites (for example, leaf clipping, which can be used to attract attention during courtship or before displaying, or during play). Also, Japanese macaques have some intersite variability in courtship gestures.[14]

In surveying the cultural primatology literature, I was struck by the absence of reports of social conventions, especially compared with the relative wealth of data on differences in subsistence techniques. If I had not known that capuchins have at least a few intriguing social conventions, I would have thought that social conventions were a uniquely human trait. However, they clearly are not. Even in the absence of language and lacking the capacity to perform complex symbolic manipulations, animals are capable of devising rituals for the purpose of communicating about their social relationships, and of transmitting those traditions to others. But why are such conventions not more commonly reported in animals? Probably such traditions are valuable only to animals living in relatively complex societies with long-term relationships, shifting alliances, and the need to coordinate with one another in cooperative activities—but capuchins and humans are certainly not the only species that meet these criteria. Could it be that many other animals create novel bond-testing rituals, as well, but these have simply been overlooked because no one has specifically searched for them or collected data in a way that permits these behavior patterns to be noticed?

To produce convincing data sets for the study of traditions (particularly the development, social transmission, and geographic distribution of social conventions) requires years of data collection and collaboration between researchers working at different sites. It is no wonder that such data sets are scarce, and I am immeasurably grateful to my fellow capuchinologists for being so collegial and patient as to cooperate with me in this large-scale research project. We hope that researchers will become interested enough in these issues to study them in other species as well.

# CHAPTER 12

# Social Learning and the Roots of Culture

The human capacity for culture is a large part of what sets us apart from other animals, and the question of how this capacity emerged is one of the great puzzles facing biologists and social scientists. Whether culture is exclusively human is largely a definitional quarrel. Anthropologists have formulated literally hundreds of definitions of culture, many of which exclude nonhuman animals either explicitly or by including language as an attribute of culture or the means by which it is transmitted, or both. A minimal—and, in my view, more useful—definition of *culture* is "behavioral variation, the distribution of which is determined at least in part by social learning," with *social learning* defined simply as "use of public information to organize behavior."[1] *Individual learning* means learning on one's own, on the basis of direct experience with a problem or task (for example, through trial and error, insight, or recombining skills). And of course some behavior is partially unlearned or "innate."

Many species, including not only primates but also rats, bats, dolphins, and many songbirds, acquire important parts of their behavioral repertoire by social learning.[1,2,3] Why might the capacity for social learning have evolved? What causes some animal species to rely on social learning more than others? Copying what your elders or peers do isn't always the best solution to a problem. If they know more than you do, it is wise to copy them. But if your companions have less experience than you do, or different needs from yours, then you may be better off figuring out a solution on your own. And, as teenagers like to remind their parents, your elders are likely to be poor sources of useful information if conditions have changed substantially since they were young. Ideally, natural selection should have shaped many animals to be intelligent social learners, selective of the contexts in which they attend to social cues as well as to the characteristics of the models that they copy.

My initial forays into the geographic patterning of capuchin monkey behavior led me to wonder more about how and when capuchin monkeys learn from one another. Although there was a growing literature in the emerging field of cultural primatology documenting between-site variations in behaviors of various sorts (mainly foraging-related behaviors), there was virtually no detailed documentation of the types of social interactions that might bring about such patterning.

For example, it was known that chimpanzees at some sites (but not at others where nuts are also available) use hammers and anvils to crack open nuts.[4] Infants have many opportunities to watch their mothers engage in this behavior, so learning by observation of a model was a likely form of social learning. It was also possible that young animals learned socially not by watching the process of nut cracking but by observing the material remains of nut cracking and by begging for nuts that had just been cracked. Another possibility was that adults might actively assist young animals to crack nuts by modifying their hand positions. The patterning of nut cracking and various other subsistence skills across chimpanzee communities could not easily be explained by genetic or ecological variation between these sites, which lent credence to the hypothesis that such variations were "cultural" in origin. But the extent to which young chimpanzees' nut-cracking techniques were enhanced by the opportunity to observe, interact with, or scavenge from skilled adults was unknown.

It was also unknown whether, given that social learning *was* exerting an effect on technique, young chimps were more likely to copy the subtle details of hand positions and choice of materials of those individuals

(a) whom they observed most frequently, (b) with whom they had the closest and most tolerant relationships, or (c) who were the most successful at nut cracking. There were many other unanswered questions as well. Was there a critical period in development when it was easier to learn to crack nuts and when social learning was more effective? Was the mesh of social, motor, and social capabilities shifting with development so as to periodically open and close different avenues of social learning, as has been suggested for orangutans?[5] Did animals attend to social cues more often during nut cracking than during other types of activities, and if so, why?

Although these sorts of questions had obvious importance for a thorough understanding of "culture in the wild," they were difficult questions to answer, because they necessitated the cooperation of large numbers of scientists who could patiently follow the same animals for the many years that it took them to attain adulthood. A project of such magnitude was simply not feasible, given the time constraints of university teaching obligations and the limitations of funding. In the United States, getting a research grant that lasts longer than three years is almost impossible, and if obtained, such grants in the behavioral sciences are typically insufficient to fund large teams of researchers.

As inconceivable as such a project might be for scientists in the United States, it proved to be a perfectly reasonable prospect in Germany, where basic scientific research has a high funding priority and the general public is tremendously enthusiastic about questions of human origins. In 2001 I was offered the opportunity to head a research group at the Max Planck Institute for Evolutionary Anthropology, to investigate the evolution of culture by examining cultural propensities in animals other than apes and humans. It was a five-year position that came with virtually unlimited funding for equipment and for the salaries of field assistants, data analysts, and graduate students. Although it meant moving my family overseas and having to learn a new language, I couldn't pass up the opportunity to do a developmental study that would answer such questions.

Nonetheless, I was going need a lot of help to make the appointment work. UCLA, where we were teaching, would not give us more than a three-year leave of absence, nor would the university let us leave the campus for long trips during the first year of my appointment in Germany. Plus, I would have to train several new field assistants in Costa Rica while simultaneously teaching a course at UCLA. It didn't seem possible.

And it would not have been possible, had it not been for the heroic efforts of Hannah Gilkenson. Hannah was an extraordinarily talented field assistant who joined our project in January 2001, during a phase when matters were not going particularly well. The stress of caring for our young daughter (often without the help of day care) while working against the ticking tenure clock at UCLA had led Joe and me to be somewhat less careful when selecting graduate students and field assistants to work with us in Costa Rica that year, and we also had no time to micromanage the project staff and anticipate problems before they became crises. The result was the least pleasant field season in the project's history. Some of the newer project members were so busy partying, yielding to their raging hormones, and meddling in other people's personal lives (even snooping in others' private journals, mail, and computer files to find ammunition for manipulating their coworkers) that life in the project house was sheer hell, particularly for those of us who believed in the importance of moral conduct.

The experience made us long for the old days when we had a smaller and more amiable staff, and we seriously doubted the wisdom of expanding the project to include as many field assistants as would be necessary to conduct a developmental study of multiple monkey groups. But awful experiences often have a silver lining, and the saving grace of the 2001 field season was the discovery that Hannah was not only a superb fieldworker but also a mastermind at dealing with unpleasant social situations and at creating order from chaos. I would not have taken on the job in Germany if Hannah had not agreed to supervise the field site in my absence.

Hannah was just twenty-three years old when she accepted my job offer, and I am still amazed in retrospect that I entrusted such a big responsibility to someone so young—but I have no regrets. At the end of that field season I headed off to UCLA and then Germany, leaving Hannah to learn Spanish, manage a $50,000 annual budget and keep good enough records of all financial transactions to satisfy German accountants, habituate at least one more monkey group, train six brand-new field assistants who had no prior experience, and—most important—restore the culture of the project to what it had been prior to 2001: a group of dedicated, honest, cooperative researchers who enjoyed one another's company and knew how to blend work and fun while doing high-quality research.

When I returned to the field site six months later, in January 2002, I had very little cleaning up to do. Hannah had become virtually fluent in Spanish, purchased a new project vehicle that was far superior to the old one, and rented an abandoned two-story, tree-shaded house (Yellow House) that was spacious enough to accommodate our newly expanded staff. The field assistants (all of whom Hannah had helped to select) were a fabulous bunch of people, and their terrific work ethic and cooperative nature set just the right tone for the rest of the "baby project" (as the assistants affectionately called the developmental study).

Hannah originally intended to stay for just a year, until she had the project back on its feet, but by the time she had accomplished her mission, she was so attached to the monkeys, to certain project personnel, and to the "monkey-project culture," which she had in large part created, that she signed a contract to stay on as manager until 2006. She had gone native, much to her mother's distress and our delight, and it appeared that she would never turn back. Joe and I were tremendously relieved, as by this time we could not imagine running the project without her. With Hannah in firm control of all the logistical details and personnel issues, we were able to remove ourselves from the project a bit (to the house next door, to be precise) and devote ourselves to the scientific aspects of the project and to raising our daughter.

Hannah has an inexhaustible memory for consumer bargains and could wring an almost embarrassingly high quality of life for the staff out of a relatively small amount of money. Her social intelligence is also off the charts; she knew exactly when a project member had a problem and could almost always intuit whether or when to intervene to help solve it. This was especially helpful, as we had long since forgotten what it felt like to be twenty-two years old and did not feel well equipped to provide much counseling to upset field assistants, even though we wanted to help out. It is utterly exhausting to live in cramped quarters with a group of coworkers and employees, day and night for years on end, and we remain in awe of Hannah's emotional stamina.

With seemingly infinite resources at my disposal, the world's best field manager overseeing logistics, and an international crew of cream-of-the-crop field assistants (recruited via the Internet and interviewed on the phone after extensive background checks), I made it my goal to carry out the most detailed developmental study of a wild primate species to date. First, I needed to augment my sample size with more social groups. I knew I could not count on all of the animals born in my initial study groups to survive to adulthood. Also, studying multiple troops would give me a better chance of documenting group-specific ways of conducting social interactions. Human cultural groups exhibit tremendous variation in the standards of acceptable social-interaction patterns, including appropriate relations between the sexes and between individuals of disparate dominance ranks, and appropriate ways to mediate resource distribution and to resolve conflicts. I was curious about whether such variation existed in nonhuman primates as well. The most logical group to habituate next was Pelon group, an exceedingly large group that had a home range broadly overlapping that of Abby's group. The Pelon group had been terrorizing Abby's group for the past two years, and it seemed clear that the two groups would soon be exchanging migrant males. We censused all the other groups in the area and confirmed that this was, indeed, the best group to habituate. While the habituation process was begun with Pelon group, we began collecting focal data on the recently born infants in Abby's and Rambo's groups and waited for new infants to be born.

Because this was my chance of a lifetime to learn about the development of many kinds of behavior, the data collection protocol was extremely complicated—so complex that it took about three months for new field assistants to pass all of their interobserver reliability tests. I wanted to know about motor development and the fine details of foraging techniques so I could detect any social influences on food-processing techniques. I wanted to record every nuance of gestural and vocal communication so I could find out whether there are group- or clique-specific ways in which the monkeys communicate with one another.

Finally, I wanted to determine the extent to which young monkeys rely on social cues to decide what plant species are food and when to categorize each animal species encountered as prey, predator, or feeding competitor.

To accomplish all this, I had to create codes for hundreds of plant and animal species, behaviors, and monkey identities, as well as numerous syntax markers to denote what type of data was being recorded. By the time I had finished the coding scheme, we had approximately 780 codes that had to be memorized by all assistants. Hannah plastered the walls of the project house with creative and humorous signs to help field assistants remember the trickier aspects of the data collection protocol. Memorizing the codes was actually the easiest part of the job; most of our field assistants could manage this in less than a week. Learning to apply them to the correct situations in real time was, of course, much trickier. The field assistants had to be able to reliably identify more than a hundred individual monkeys, as well as the plants and other animals living in the reserve. They also had to learn to correctly categorize monkey vocalizations.

The least favorite part of training for most assistants is the monthly vocalization test, when we play back prerecorded monkey calls during dinner and require the assistants to write down the corresponding codes, thus revealing whose data are currently reliable for the various call types. This is the one aspect of the training program that some field assistants never pass entirely. Everyone becomes sufficiently familiar with the more common call types, but rarer vocalizations are too difficult for some people to remember. Knowing this, we try to pair novice observers with more proficient observers, and we keep track of which data sets can be used for analyses of vocalizations and which cannot.

Data collection is difficult even for a field assistant who knows all the codes, monkeys, and plants, because the monkeys spend much of their time moving rapidly across difficult terrain. For this reason, we collect data in teams, so one person can keep his or her eyes fixed on the focal animal at all times while the other records the data or moves to get a better view. The first person narrates aloud while the second observer types the codes into a handheld computer and double-checks what the narrator is saying. Both observers double-check monkey and plant identifications throughout the course of the focal follow, and if there are disagreements as to the identity of a monkey or a sequence of actions, then the data are discarded and a new follow is begun. Disagreements about plant identifications are resolved by collecting a specimen and bringing it home to check with an expert, and sometimes a vocalization can be checked by recording the sound on a microcassette and later comparing the recording with those on the vocal-repertoire CD.

One of the most difficult aspects of data collecting is typing quickly enough to catch all of the action. Before field assistants are allowed to collect real data, they practice speed typing by transcribing tapes that Hannah and I have made of fast-action focal follows. Once they have mastered this skill to the point that they no longer miss codes or omit lines of data, they begin "shadowing" teams of trained observers in the field, and we compare their transcription to the official transcription of the data at the end of the day. It is much harder to type when you are hanging off a cliff, crashing through brush, or wading through a river than when you are lying in the hammock at Yellow House. Trainees are ready to collect real data only when their "shadow" follows reliably produce data that are identical to the data collected by the trained observers. Of course, the monkeys occasionally foil even the most experienced observers by performing novel behaviors for which there are no codes, or by engaging in high-speed activities involving dozens of participants; in these situations, one or both observers whip out their tape recorders and babble away like sports announcers until the action has died down and they can resort to typing once more. These taped inserts are transcribed and inserted into the data the following day.

With so many people working on our project (up to twelve at a time during the phase funded by the Max Planck Institute), it would be easy for mutations in the data-collection protocol to arise and be propagated, as in a children's "telephone" game. By regularly shuffling people among the different data-collection teams, and by conducting monthly code tests, we make sure that little or no "drift" occurs in the way the data are collected. Every night at what we call "Monkey Notes," during dinner, we discuss any disagreements or doubts that have arisen during the day's data collection so that the problems can be resolved before the data are edited and the following day's data are collected. One problem stemming from our highly international community of field assistants (which has included, at various times, people from Sweden, France, Scotland,

Germany, Costa Rica, and the United States) is that we often have trouble understanding one another's accents and vocabulary. This problem is amplified when we are trying to yell to one another over the noise of rain, wind, parrots, cicadas, or radio static. Speaking "monkey Spanglish" helps us avoid such misunderstandings—we all use a highly standardized set of phrases for describing the monkeys' actions, incorporating the easiest-to-understand elements of English and Spanish phrases, and codes, in a strict grammatical sequence. This system of communication has evolved over time as we have figured out which ways of describing particular behavioral sequences are least likely to be confused with other things we might say in similar contexts.

When we were writing this book, the developmental study was just about to start its fifth year, and we were still cleaning the 2004 data set; the project is still very much a work in progress. Food processing was the first aspect of development that I began to analyze. In the cross-site study of food-processing techniques that I conducted earlier with my collaborators from Palo Verde (Melissa Panger) and Santa Rosa (primarily Lisa Rose), we uncovered many differences in the ways foods were processed.[6] As I have noted, the ecology of the three sites is similar, and the capuchin populations probably do not differ substantially in their genetic composition because they were quite recently linked by forest corridors. So it is quite likely that the intersite differences we identified represent divergent social traditions.

In a few cases—behaviors such as wrapping leaves around hairy caterpillars or hairy *Sloanea* fruits to avoid being pricked—the technique was used by just one or a few individuals, and thus the trait appeared to be an innovation that had not spread far. Most of the twenty cases of site-specific food-processing techniques consisted of applying a common element of the general capuchin behavioral repertoire (such as pounding, scrubbing, or using a fulcrum) to a different situation. For example, *Cecropia* fruits were routinely pounded and scrubbed by monkeys at Santa Rosa, whereas they were not processed before ingestion at the other sites. When a food required processing prior to ingestion, there was usually also some within-site variation in the way it was processed.

But were the intersite differences we identified really cultural in nature? One way to address this question was to check whether, within each social group, individuals that associated and observed each other's foraging activity frequently also shared the same techniques. If this were the case, we would be one step closer to being able to confidently say that monkeys adopt particular techniques because of social influences— that their behavior is, in a sense, "cultural." And indeed, when we performed these analyses, we discovered that for several foods, those pairs who shared the same technique also spent more time together.

I decided to take the question a step further by looking more closely at the sorts of social interactions that might lead to social learning of foraging tactics, and by looking at the developmental processes that lead to acquisition of complex feeding techniques. Capuchin monkeys often exhibit a behavior that we call food interest, in which they observe another monkey's foraging behavior at close range. The following observation offers a typical example:

> Yoyo, a juvenile male, is foraging on *Bromelia* fruits. He stands on a stalk bearing several yellow fruits and bites the stem, tugging with his entire upper body. After trying several angles, he breaks off a cluster of fruits, creeps carefully out of the plant, treading on the centers of two bromeliad leaves so as not to get pricked by the spines on the edges of the leaves, and scampers up a tree trunk to eat peacefully on a branch. He tilts his head back and holds the fruits above his head, gnawing vigorously on the tough fruit until it cracks open and the juicy white pulp starts to ooze out. Then he squeezes the fruit through his teeth, just as human children squeeze *gelatinas* [frozen Jell-o] through holes in plastic bags and into their mouths. Eldritch, a yearling female, hurries over to see what Yoyo is doing. When she is just a foot away, she stops running and cautiously approaches, sticking her face into the cluster of fruits. She stares intently at Yoyo's hands and mouth from a distance of three inches and watches mesmerized as he squeezes more pulp into his mouth. Tentatively, she touches another of the fruits in the cluster and then looks back at Yoyo's mouth. She trills and touches Yoyo's mouth as he begins chewing on a second fruit. Yoyo ignores Eldritch but tolerates her investigations. Finally Yoyo finishes his fruit and drops the cluster of depleted fruits on the branch before returning to the bromeliad for more. Eldritch briefly inspects the abandoned fruits, licks some juice off of the empty husks, and follows Yoyo.

It is often hard to interpret these interactions. In more than 80 percent of food-interest interactions, the observer is an immature animal

Infant Dashiki watches his uncle, juvenile male Hada, forage on *Bromelia* fruits.
Photo: Susan Perry.

rather than an adult. What are the youngsters getting out of these inter-actions? Are they trying to learn from more experienced foragers? If so, what are they learning—what to eat, or how to process the food? Or are they merely begging and not trying to learn anything at all? Although most food-interest interactions do not result in food transfer, some do, and perhaps this moderate frequency of reward is enough to keep the animals interested. I decided to rank all of the items in the capuchin diet according to rarity (that is, the frequency with which they are eaten by other group members), difficulty in processing (the number of processing steps necessary before the food can be ingested), and size. Then I examined whether these factors affected the propensity of the monkeys to exhibit food-interest to groupmates foraging on these foods.[7] Young monkeys were far more likely to show interest in rare foods than in common foods, which suggests that they may be trying to find out more about novel dietary items. Monkeys were also more interested in watching foraging on larger food items, which is consistent with the idea that they may be begging for leftovers. Perhaps the most intriguing result was the finding that the more steps that were necessary to process the food, the more interested they were in watching others forage on the food, even when handling time was taken into account. This suggests that youngsters who observe foragers may be trying to learn something about techniques used to process food. This result was more robust than the result that implied that the monkeys were begging.

Now I had three forms of evidence suggesting that social influence guides capuchins' decisions about how to process food: processing techniques differed among study sites; within each troop, frequent associates tended to share the same processing techniques; and food interest was elicited most often by consumption of food requiring the most elaborate processing. What puzzled me was the fact that all available evidence from experiments done with captive capuchins suggested that they do *not* copy the techniques used by their peers to process food. For example, Elisabetta Visalberghi's research group had given brown capuchins the opportunity to watch a model monkey push a peanut out of a tube with a stick.[8] Even after extensive opportunity to watch the model perform the task correctly (up to seventy-five times), when they were presented with the same task, the observer monkeys failed to copy the model's actions in order to obtain the reward.

Social learning studies in captive capuchins suggest that social influence is important in shaping many aspects of feeding behavior (mainly food choice), but that capuchins do not imitate; that is, they do not copy the motor patterns used by groupmates. Rather, they end up learning similar skills and having similar diets because of their tendency to coordinate their behavior with one another. Because they are in the same place as their groupmates at the same time, coordinating activities, they end up being exposed to the same types of stimuli and subsequently learn similar solutions to problems via trial and error.[9]

The "baby project" offered a unique opportunity to help resolve the controversy over social learning in capuchins. I chose a set of foraging tasks of varying difficulty and set out to trace the natural acquisition of skills in the wild; we would document opportunities for social learning as well as changes in individuals' techniques over time. This sort of study would bring a life-history perspective to the topic, and it would help us figure out how the nature and difficulty of a task affected the monkeys' probability of relying on social cues. We would be able to document how the animals' motivation to seek information from members of their own species changed over time as a function of their skill level, and we could also see whether they changed association patterns in accordance with age, skill level, and so on. During the focal follows of infants, we began recording the activities of all animals within five body lengths of the focal so we could see which activities triggered the most visual attention by infants.

Because our previous work had shown that visual attention increased with a food's level of processing difficulty, I chose four challenging foraging tasks—three fruits and one spiny caterpillar—as primary foci of study. When animals began foraging on *Luehea*, *Sloanea*, or *Sterculia* fruits, all of which require multiple steps to open, we stopped focal sampling and switched to a special protocol. We did our best to record the techniques used by each monkey foraging on these items. Every time a monkey picked another fruit, we recorded in minute detail the actions of each hand and the mouth until the fruit was dropped. We also noted the distances of all other foragers to the focal animal, and the gaze orientations of all foragers. Then we moved on to another forager, rotating through the group until the last forager had finished eating.

It is May 2005, just after a typically busy day in the field, and we are all assembling for group dinner. Usually Joe, Kate, and I eat in our own small house next to Yellow House, but once every seven to ten days, we join Hannah and the assistants for supper and I lead a seminarlike discussion of some topic in primate behavior. I have been at home analyzing data today, in preparation for a talk I will give in a couple of weeks at a conference on social learning, so tonight we are going to discuss project data rather than a journal article. I clamber over the wall in the yard between the two project houses to join Joe, Kate, and the others, bearing my contribution to the meal: the chutneys (green tomato and mango-pineapple-ginger) I had simmering on the stove all afternoon while I was working on the computer.

As I enter the project house, my senses are bombarded by the usual evening array of sounds and smells: the Dave Matthews Band is blaring on the boom box, and field assistant James Broesch is singing along as he downloads data from the handheld computers to the laptops. Tantalizing aromas from Colleen Gault's dinner preparations are wafting through the house, mercifully masking the smells of soggy jungle boots, sweaty field clothes, and wet dogs. Several other field assistants are zipping about the project house, dodging in and out of the shower, packing their lunches for tomorrow, and assembling their gear. Hannah is sitting at her desk, attempting to get some receipts in order. Six-year-old Kate is shrieking happily and running around the room with her well-worn Piglet doll, dragging behind her a two-meter length of hot pink flagging tape. Two kittens are chasing Kate and pouncing and batting at the fluttering tape.

The dinner table is already heaped with food: *channa masala*, curried chicken, *raita* made with Colleen's homemade yogurt, and some vegetables so thickly coated in spices that it is difficult to recognize the species. Freshly cut, unadulterated pineapple lies on a platter at one end of the table for those who are scared of the spicy stuff, alongside Colleen's homemade *naan* (Indian bread), which is a favorite with everyone. When I head into the kitchen to get my plate, I find the Tico team members gingerly picking the cardamom pods, cinnamon sticks, cloves, peppercorns, and bay leaves out of their rice and flinging them back into the rice cooker. We don't all share the same culinary preferences, but there is no doubt about it: we eat well on this project. And because the cooking duties are rotated, everyone gets what they like some of the time. Personally, I am in heaven whenever Colleen cooks, and as I heap my plate, I feel very sorry for my colleagues in Africa and Southeast Asia who are toughing it out in the wild doing "real" fieldwork, and hence having to eat plain rice and sardines for dinner every night. Here in Bagaces, why suffer when we don't have to?

I cautiously turn down the volume on the Dave Matthews Band, and when I don't receive a reprimand, I sit down at the table. Everyone is already ravenously tucking in to their dinner, hoping to devour it quickly enough to be ready to compete for the surplus *naan*.

"Let's get started," I say. "Tonight we're breaking tradition and just talking about Lomas data. But it ties in with what we have been reading in the Richerson and Boyd book, the Whiten papers, and the Ottoni paper.[10] For this upcoming social learning conference, I'll be analyzing the *Luehea* data. The *Luehea* fruit makes a nice comparison with Whiten's artificial fruit studies, because there are basically just two ways to open it. But of course the animals in our study will have unlimited opportunities to learn the task since we're following them for their entire lives, so we won't have the problem most lab researchers have, of wondering whether the monkeys have had enough time to figure out the solution. Also, since *Luehea* is such an important part of the diet, I think we can be pretty confident that just about everyone is sufficiently motivated to learn the task. When Juanca and I analyzed the 1992 diet data last year, we found that *Luehea* was 4.4 percent of the overall diet and up to 15 percent of the diet in some months. It was available for about half of the year, so it was clearly an important backup food even when it wasn't the preferred food on the daily menu."

"You're not going to look at *Sloanea* processing? After all the times we've had to get those nasty hairs extracted from inside our eyelids? You'd better use those data!" says Cindy Carlson, laughing.

"Yes, yes," I hastily reassure her. "There will probably be more interesting patterns there, since there is so much variability, but for this brief conference talk I thought I would do something that is simpler to analyze. *Luehea* is great because there are just two techniques used, pounding and scrubbing. It's easy for the audience to follow the analysis. I'll get to the *Sloanea*, though—we didn't do all that suffering for nothing."

"Hey, wait a minute," says James. "Before we get too far into this, do you want to see the video clips I found when I was backing up the video footage onto DVDs? There is awesome footage of Power pounding, and there is some pretty good footage of scrubbing too. I think you'll want some of this for your talk." James, a former computer consultant, has been working on cleaning up our video database while his back heals from an injury.

"Great! Let's see it," I say, and James gets the laptop and turns the screen to face the group. "Power first," he says. The image zooms in on adult male Power's hand, holding a *Luehea* fruit. The large woody capsule has five cracks in it. Inside these cracks, invisible to the camera, are tiny, nutritious seeds. Eventually the fruits will dry and the cracks will open, releasing these tiny seeds to the wind for dispersal, but the monkeys attempt to open the fruits when the seeds are still firmly lodged inside. Power is sitting on a rock, holding the fruit in his left hand with the stem between his thumb and forefinger. He rapidly and energetically pounds the fruit on the rock, cracks facing downward, with his right hand positioned to catch the seeds as they fly out of the cracks. Every few seconds he dips his face down to eat the seeds from the fruit, his hand, and the rock. When those seeds are gone, he repeats the pounding process. His actions are highly stereotyped, rapid, and efficient.

"OK," says James, introducing the next images, "all the scrubbing ones were backlit, but here is one that's pretty good of Celeste." The camera zooms in on an adult female, sitting on a branch with her tail wrapped around a fork in the tree. She holds a *Luehea* fruit in her right hand and scrubs it rhythmically back and forth along the branch, the friction of the rough bark against the fruit tip dislodging seeds from their cracks. She keeps her left hand positioned just below the fruit as it moves, ready to catch the falling seeds. Periodically she lifts the fruit to her lips and plucks out any protruding seeds with her tongue and lips, sometimes using her fingers too.

"Scrubbing just looks so much less efficient," says Marie Kay.

"It really does," I agree. "I used to think that the scrubbers were real morons. But it turns out to be no less efficient than pounding."

"So you analyzed our experiment from the other day?" asks Susie Herbert eagerly.

"Yes, I did," I say, and I turn to explain this remark to the rest of the group. "The other day, Susie and I were discussing this issue of foraging efficiency in the field. The problem is, as you know, these seeds are so tiny and the monkeys are so high up that it's impossible for us to tell how many seeds are dropped and how many make it into the monkeys' mouths. Also, you can't tell how many seeds are left in the fruit. Anyway, the monkeys were being uncooperative on Trail to Somewhere, and so Susie and I decided to use our time more efficiently by trying to process *Luehea* ourselves and counting up how many seeds each technique yielded. We collected all the *Luehea* fruits we could find, and we picked pairs that seemed matched for openness of the cracks and number of seeds showing and put a member of each matched pair into a separate pile. Then I processed one pile by scrubbing and the other by pounding. Each fruit received ten seconds of processing."

"So Susan was the monkey and you were the observer?" asks Juanca, for clarification.

"Yes," laughs Susie. "I timed her with the chronometer, we counted the seeds, and I wrote it in the field notebook. She was doing it just like a monkey, except that she didn't eat them."

"Pounding yielded 7.8 seeds and scrubbing got me 5.8 seeds for the same amount of processing time," I say. "There is no statistical difference."

"That's really pretty cool," says Susie thoughtfully, "because if there is no advantage in efficiency to one technique versus the other, then you wouldn't expect trial-and-error learning to bias the monkeys toward using one of the two techniques."

"That's right—so their choice of technique is more likely to be swayed by social cues than would be the case if one technique were better than the other," I say. "OK, so here is the distribution of techniques across groups." I hold up a scrawled graph. "Pounding was the predominant technique in both groups (about 75 percent of all processors), particularly for males. But each social group had strikingly different proportions of pounders versus scrubbers. In Rambo's group, all of the mature females were pounders, whereas in Abby's group about three-quarters of the females were scrubbers. Pelon group had a more even mixture of techniques."

"But don't some monkeys do both?" asks Daphné Kerhoas-Essens.

"Yes, about 5 percent of adults exhibit multiple techniques, as I'll get to in a bit, and of course the young animals do all sorts of crazy things before they settle into a groove. This graph is just talking about adults though," I say. "So about 95 percent of the adults had an efficient and stereotyped way of processing in which there was virtually no deviation in hand position, type of actions used, and the rhythm of the process from one processing event to the next."

"No kidding," says Marie. "The other day I was transcribing data on Moth and it was 'Moth pounds with left hand, catches with right hand and ingests' about fifty times in a row. Adults are so boring compared to kids."

"Some of the kids are hilarious though," adds Colleen. "They are so creative and so inefficient. There's no way we could code that stuff—I guess that's why you make us do it all on microcassette recorder, huh?"

"That's right," I say. "I worry that we'll straightjacket some of the more creative stuff into preexisting codes if I don't give you the freedom to describe this stuff on the fly."

"Wembley is funny," says Marie. "Have you guys been noticing what he does?"

"I've got some good video of Wembley right here," says James.

"How old is Wembley?" asks Cindy, who hasn't worked with Rambo's group.

"Three years old," says Hannah.

We stare at the screen as Wembley selects a *Luehea* fruit and inspects the tip of it for loose seeds. Then he bites the stem connecting the fruit to the tree and pulls it off. He backs down the branch and rolls the fruit between his hands and the branch, sort of juggling it between hands and branch in an uncoordinated way. He nibbles on the tip of the fruit again, but his technique has been completely ineffective; no seeds have been loosened. Once again he scoots backward down the branch, this time dragging the fruit on its side. This time a couple of seeds are loosened to the point that he can pluck them from the fruit. Once again he walks backward, this time holding the fruit loosely and sort of dragging it along. He scrubs it sideways with his left hand twice and then abandons the fruit.

"Do you have any data on Rain from 2001?" asks Hannah. "He was really funny, as I recall."

"Yes I do, but not on video," I say, as I reach over to call up the file. "Here you go: this is from when he was eighteen months old. 'Rain scrubs the tip of his *Luehea* fruit against the branch, holding it in a lop-sided manner, and then pounces on his *Luehea* fruit and bats at it. Then he grasps the stem in one hand and flails it about, occasionally banging it against the branch. Next he rolls it against the branch. Then he takes it in two hands and bounces up and down, using the motions he normally uses to splash water out of a tree hole. Then he holds the fruit in his right hand and bats it with the left hand while bouncing up and down. None of this is particularly effective, as the few seeds that he does dislodge flutter away in the wind, and he finally abandons the fruit.'"

"When exactly do they start processing?" asks Daphné. "I can't remember who was the youngest one I've seen doing it. I mean really processing, not just eating stuff from the ends."

"During the first year of life, you're right, most infants just inspect the tips of *Luehea* fruits and try to get what they can to eat without any fancy processing. But they exhibit a great deal of interest in how others are processing *Luehea*. During their second year of life, they are already expending a great deal of time and energy attempting to extract *Luehea* seeds from deep inside the fruits, and they use a wide variety of techniques. By approximately three years of age (ranging from one to five years), pretty much everyone settles into a single technique that they use more than 75 percent of the time in a highly standardized way.

"So here is where it starts to get interesting: we found that about 70 percent of all monkeys for which maternity and maternal processing style were known matched their mother's technique, and this was statistically significant."

"I wonder whether you could separate out genetic influence by looking to see whether they were just as likely to match their fathers," says Susie. "Presumably they spend less time with dads than with mums."

"That's a great suggestion," I say. "Like you say, offspring spend about seven times as much time with mom as they do with dad. If this matching is due to social learning rather than inheritance of a genetic disposition to prefer one technique over another, you would expect them to match mom far more often than they match dad. I did try to look at that, but it was a bit of a mess to interpret. Eighty-five percent of the monkeys matched their dad, yet this was not statistically significant. The problem is, all but one of the dads were pounders, and pounding is the predominant technique in the population generally. Only one dad was a scrubber, and he had only one offspring. So the data are basically uninterpretable. Perhaps some day more scrubbing males will migrate and breed and we can get a decent sample."

"What about prestige-bias transmission?" asks James. "Did you check to see whether they are more likely to watch the more skilled processors? That is so common in humans; I wonder whether capuchins do it too."

"No, I couldn't really do that. The problem is, I can't reliably assess which monkeys are the best *Luehea* processors, for the same reason I couldn't tell which technique was better just by observation. That Ottoni et al. paper we read last week was the first nonhuman primate paper I've seen that seemed suggestive of prestige-bias transmission.[10] He was looking at nut-cracking abilities in brown capuchins and found that young animals were far more likely to observe the monkeys who were more successful nut crackers. It wasn't entirely clear whether the monkeys understood the concept of proficiency or were just scrounging from the monkeys that produced the most food, but it amounts to the same thing in terms of observation opportunity: whether they are begging or trying to learn, they end up watching the most proficient models. I wish I could do similar analyses for our monkeys, but there aren't any tasks except for *Sterculia* processing for which you can assign a reliable proficiency ranking, and we don't yet have a big enough sample size for *Sterculia*."

"Did you look at their observations of anyone besides mom?" asks Colleen. "It seems that once they are off mom's back, they don't spend all that much time around her, and other monkeys might be more important models."

"That's absolutely right," I answer. "Because infant monkeys start exploring on their own and making friends beginning around three months of age, there are many other potentially important social influences besides the mother, and each juvenile has a distinct social network. So I decided to examine years two through four of development more carefully, noting how many times at that age they had a chance to observe pounding and scrubbing when they were foraging on *Luehea*. The monkeys tended to adopt the technique to which they had greatest exposure (though this was statistically significant only in years two and four), and by the time they were in their fourth *Luehea*-processing season, all of the monkeys in the developmental study had conformed to the technique they had observed most often. The most interesting cases were those in which juveniles initially settled into one stable technique that was highly effective and later switched. Both Cassie and Marañon were scrubbers just like their mothers during their second and third years of life, but they began experimenting more with pounding during year four of their development, when they were exposed mainly to pounding. By year five, both of them had become stable pounders."

"Weren't you saying in the forest the other day that the females tended to do what mom did and the boys were more prone to influence from monkeys other than mom?" asks Susie. "Kind of like in that *Nature* paper we read about the six juvenile chimps, in which the three females learned quicker and made tools that were more similar to those of the mothers than the tools the males made?"[11]

"Oh, that didn't hold up when I threw in more social groups," I answer. "That was the way the results were headed when I had analyzed just the twenty-seven or so kids from Abby's group, and there is a small tendency in that direction still, but it's not significant with the larger sample."

"Too bad—that was a fun result. Do you know yet whether females learn faster than males?" asks Susie.

"I haven't tried to analyze it in a very precise way yet, but just rounding to the nearest year, it looks like females' techniques stabilize about six months earlier than males' do. Of course stabilization isn't quite the same thing as competence. One thing I want to do eventually, when everyone in the sample has stabilized, is to see whether the total amount of social exposure to other *Luehea* foragers affects the age of competence or stabilization of technique. One of the most prevalent ideas in the primate social learning literature is that monkeys can learn effectively from one another only if they are frequently exposed to individuals who tolerate their close physical proximity in a relaxed setting.[12,13] If this is true, then we would expect variation in monkeys' abilities to learn socially, with the more gregarious and popular monkeys conforming to the group's most common techniques early in life and the peripheral, unpopular monkeys exhibiting more eccentric techniques until late in development."

"*Como* Ripley" ["Like Ripley"], says Juanca, and everyone who works with Rambo's group bursts out laughing. Ripley is a highly paranoid, neurotic monkey who is very rarely seen in the center of the group and almost never socializes with others. In fact, Rambo's group had been studied for nearly eight months before we even met this female. When I first spotted her and was describing her appearance in the field notes, wondering whether this would be our first confirmed case of female immigration, a two-year-old juvenile whom we had assumed to be an orphan astonished me by running up to her and nursing. This was why I decided to name her Ripley, after the TV program *Ripley's Believe It or Not*. Even now that Ripley has been accustomed to the daily presence of human observers for nine years, she is rarely seen in the center of her group and has very few social contacts. One of the many unbelievable

aspects of Ripley's behavior is her astounding incompetence at processing *Luehea*. Her inadequacy at *Luehea* processing is a favorite conversational topic, and everyone starts imitating bizarre methods they have
seen her try. Alex starts belly laughing just thinking about Ripley.

"Last time I saw her processing, she held the fruit with the cracks facing upward—you know, the opposite of the way in which most monkeys
hold the fruit—so that the seeds couldn't even fall out of the cracks, and
she was tapping the side of the fruit," says Marie.

"Did she get anything to eat?" I ask.

"Maybe one seed."

"I've seen her hold it point downward in her right hand and tap it
with the left hand," says Susie.

"Several times I've seen her do a U-shaped swinging motion in which
she whacks the edge of the fruit against a branch on the upswing. She always seems to drop the fruit without getting any seeds to eat," I say. "It is
tempting to think that her total failure to learn a skill that 95 percent of
the population masters by age five is due to her lack of observation opportunity. She's always alone, aside from a nursing infant, when I've seen
her try to eat *Luehea*. She acts the way a yearling or two-year-old does,
the way she uses a wide variety of ineffective techniques rather than settling into a single pattern. But we don't know anything about Ripley's
'childhood,' of course, and there are other loner monkeys who are perfectly competent. We'll have to wait till our developmental cohort has
matured to really tackle this problem."

"Speaking of bizarre techniques, I can't believe you haven't mentioned foot pounding yet," says Hannah. "That has got to be the funniest *Luehea* technique ever."

"Foot pounding? What's that about?" asks Daphné.

"Oh, you've got to see it. It's so funny," says Cindy. "Jester always does
it. She holds the fruit in her foot to pound it. It loosens the seeds just
fine, but since Jester uses both hands to hold on to the branch while
pounding with her foot, neither hand is free to catch the seeds that are
dislodged from the fruit by the pounding action. She does manage to get
a little food this way though. I've seen her transfer the fruit to her hands
to inspect the tip and pull seeds out of the cracks after they are moved
forward. I've been trying to get this on video, but Flakes hardly have any
*Luehea* trees in their range."

"It's such a shame, isn't it?" I say. "Before their range shifted, they ate quite a lot of *Luehea*. Foot pounding is an innovation created by Jester when she was a low-ranking and peripheral member of Abby's group, and we first noticed it in 2001, though she may have invented it much earlier. The only other monkeys seen to foot pound are Jester's daughter Mead, her son Jade, and another peripheral female, Eldritch, who subsequently left Abby's group with Jester to form Flakes group. Mead and Jade experimented with foot pounding, but they seemed to prefer using more efficient techniques: scrubbing in Mead's case, and pounding with the hand in Jade's case. But the fact that the only animals to adopt this very strange and inefficient technique are the offspring of the innovator and one of the innovator's closest female associates makes it look suspiciously like social transmission. Most individuals don't try this even once, so far as I can tell. Apparently social learning can transmit ineffective techniques as well as effective ones, though it seems that trial-and-error learning eventually weeded out foot pounding from the repertoires of everyone but the innovator."

"Is Eldritch still doing it?" asks Hannah. "She was so funny in 2001. Remember how enthusiastically she used to foot pound?"

We call up her behavioral records for another laugh: "Eldritch pounds with the right foot, while clapping her hands together against the branch. She slaps the branch with both hands while bouncing up and down in what looked like a frenzied dance, fruit clasped in her foot."

"So far as I could tell, Eldritch never did succeed in getting any food by foot pounding," says Hannah. "She always seemed so frustrated. I wonder if she finally just gave up trying."

"So were you surprised that there was a lot of social influence, or did you expect it to turn out that way before you analyzed the data?" asks Cindy.

"I have to admit I was a little surprised that there was a statistically significant effect of social influence, since you don't often see monkeys staring directly at one another while they are foraging. It usually seems like they are totally absorbed in their own efforts, but then they will occasionally, in about 3 percent of their foraging attempts, wander over to carefully watch what someone else is doing. I've seen several instances, though, in which they stared at someone else and then returned to foraging without doing what they had just seen the model do; I wasn't even positive that they were picking up on those fine-motor details."

"So what is the punch line going to be for this talk? Is there one main message you want to convey to the other people studying social learning?" asks James.

"I guess the most striking aspect of this developmental study is not the fact that social influence is important but the fact that it operates on such a slow time scale. Researchers of social learning typically assume that social learning mechanisms such as imitation are shortcuts to learning new skills quickly. So it is something of a surprise that the monkeys take so long—up to five years—to conform to what their associates are doing. Laboratory researchers investigating social learning mechanisms have typically concluded that monkeys can't copy motor patterns exhibited by a model if they fail to do so after seventy-five trials, sometimes fewer. But our monkeys probably process *Luehea* about 1,000 to 1,500 times per year over a three-year period before they conform to what their associates are doing."

Hannah casts me a meaningful look and glances at the clock, which says 8:45 P.M. "Oops!" I say. "Sorry about that. Let's stop here and get the table cleared so we can all go to bed. We can finish this discussion tomorrow at 4:00 A.M. in the *mono*mobile, if anyone is interested."

I scramble over the wall and get home just in time to read Kate her third bedtime story, a chapter from *Wir Kinder Aus Bullerbü* (*The Children of Noisy Village*, by Astrid Lindgren), and tuck her in under her mosquito net.

"Shooboo!!" she demands, still using her babytalk term for Schubert, who has been her favorite composer since she was five months old. I hasten to put on her CD of Schubert songs—*Die Schöne Müllerin*—and then turn out the light.

# CHAPTER 13

# Nobu and *La Lucha sin Fin:* Conservation of Tropical Dry Forests

One reason I chose to establish a capuchin field site in Costa Rica was the country's internationally recognized commitment to preserving its tropical forests and promoting research in tropical biology. In 1990, when I first set foot in Costa Rica, ecotourism was booming as a result of its pro-environment policies. I also knew I wanted to establish a field site that I could maintain for several decades, and that therefore I needed to find a site that would be safe to bring children to, once we had started a family. With its very high standards of public health and its long tradition of democracy and international neutrality, Costa Rica seemed like the sort of place that I could confidently return to year after year without worrying about my family's safety or the fate of the monkeys.

Clearly the authors of the Costa Rican constitution placed a high value on nature. The constitution states, for example, that all Costa Rican citizens have the right to "a healthy and ecologically balanced environment," and that any individual has the right to denounce "any act that infringes on that right and claim . . . reparations for that deed."[1] Polluters are required by law to pay for the damage they cause, and absence of proof that a particular activity will cause damage is not considered to be a reason to postpone conservation efforts or prohibit the potentially damaging act. Costa Ricans are proud of these general pro-environment policies. In practice, however, the laws prove to be surprisingly difficult to enforce. First, they contain loopholes, and second, most citizens, including many of the people responsible for enforcing them, are unaware of their specific contents.[2]

Many environmental crimes in Costa Rica go undetected, in large part owing to insufficient resources to pay for park guards in areas that are officially protected. In addition, the generally amiable attitude of most Costa Ricans means they do not naturally meddle in their neighbors' affairs or report them if they break the law. Their aversion to conflict and their open-mindedness about alternative lifestyles are two of the most charming aspects of Costa Rican culture and help to make the country an especially pleasant place in which to live. But when these same social norms prevent people from reporting highly destructive acts to the authorities, or prevent law enforcement officials from confronting criminals and resolving conflicts, the society as a whole suffers.

**April 21, 2002**     We are in Los Angeles. I had just spent the afternoon doing my parental duty by battling Disneyland crowds with three-year-old Kate, who was celebrating a cousin's birthday. I have always been particularly averse to large crowds, shopping, commercialism, and traffic, so I had spent most of the afternoon trying to stave off a headache by visualizing my favorite parts of the forest at Lomas Barbudal.

When we arrived back home at last, I checked the answering machine and was sickened to hear Hannah's voice delivering the following message, in an uncharacteristically trembling voice:

> Hi, Susan, I hate to be the bearer of bad news, but I just don't know what to do. Nanny was killed by a poacher today. Kristen actually talked to the man who did it. She asked if he knew how the monkey died, and he said he did it. She asked him why and he said, "*Porque estaba alli*" ["Because it was there."] He was bragging that he killed her with a rock, and he claims to have killed at least twenty-two monkeys, including a mother with a baby that were just down river, which he offered to show to Kristen. We're censusing as fast as we can, but we haven't found all the groups yet; we suspect it's Jester and Jade or

Celeste and Cassie who were killed by him. The monkeys were really dispersed, fleeing from Rambo's group in an intergroup when Nanny was killed, so we're not sure who else, if anyone, was near her at the time of the killing. Nobu is just hanging around Nanny's body, cooing, and the group has left him.

Who should I call to stop this guy from doing more killing? We don't know his name, but he's been cutting down trees near Quebrada Donut for the past week, and I imagine he'll be around for a while longer. I'm afraid if we get too confrontational he'll come after us. And what should we do about Nobu? Should someone camp near the body till the group comes back? Mino said he'd do it if you want. Should we collect Nanny's body as evidence? Do we need any tissue from her, or do we already have her genotype from fecal samples? I feel so bad for Nobu. He's nursing from her dead body and he just won't leave. He's been cooing all day.

It is amazing how quickly adrenaline can make you feel thoroughly ill when you receive bad news like this. I had known Nanny for twelve years, and it was so painful to think that she had had her life snuffed out just to provide a few seconds of diversion for a poacher, for the simple reason that "she was there." And who else had been killed? Was this man going to wipe out my entire study population? I called Hannah back immediately to get the full story.

Hannah said they were not sure whether this man owned a gun, but he was apparently quite effective with slingshots and rocks thrown directly from the hand. He knew that we studied the monkeys; nonetheless, it didn't seem to have occurred to him that we cared about their survival any more than he cared about the trees he was felling to earn a living. When he was talking to Kristen, it seemed as if he was trying to show off to her by bragging about the number of animals he had killed, and he had wanted to take her on a tour of the forest to view the corpses of his other recent victims. Little did he know what an ineffective flirtation strategy this would be. As their conversation progressed, he became suspicious about Kristen's questions and changed his story, saying that actually his friend had killed Nanny, and he also changed his name and said he didn't remember where the other corpses were. (We know that a few days later, however, he told a Tico that he had been killing all sorts of animals, not just monkeys, and that he specifically bragged about killing Nanny.)

This man was bad news, and he was obviously Nanny's killer, but we were not yet sure how much other damage he had done to the wildlife of Lomas. A few days earlier our team had discovered a freshly killed, still-warm monkey from Splinter group lying on the ground, in perfect condition, as if she had just passed out and died below a mango tree by the road. In light of this new information, we were inclined to believe that she had also been the victim of a rock to the head. But as the field assistants scoured the forest over the next three days, frantically censusing the rest of our groups, we did not find any more casualties. Perhaps the man had been exaggerating to impress Kristen with his hunting prowess when he said he had killed twenty-two monkeys. Or perhaps he could not distinguish howlers from capuchins. We had found some dead howlers recently but had not inspected their bodies carefully enough to determine the cause of death.

Regardless of the exact number of killings, it was clear that something had to be done. I sent Hannah to the park service first, and got on the phone myself to everyone I could think of who might have valuable political advice. The head of the local park service branch seemed genuinely concerned and wanted to help prosecute Nanny's killer, but most of the Costa Ricans I spoke with were fairly pessimistic. We were advised to keep a low profile and not be too confrontational, if we wanted to stay safe.

Much as I valued the monkeys' safety, I was very concerned about my field crew. Most of the field assistants were women, and they were all unarmed. Many of them were outraged enough by Nanny's death to put their lives on the line to protect the monkeys, and I had to beg them to be careful and not to enact any conservation strategies I had not approved. I asked Hannah to make inquiries about arming the female assistants with pepper spray or mace, in case they had to ward off poachers who expressed excessive interest in them. But mace wouldn't do a lot of good against a man with a slingshot, if he wanted to hurt them. It was clear that our conservation efforts had to be fairly clandestine; no one wanted to suffer a fate like that of Dian Fossey. We just had to hope that our presence would deter poachers, and we had to keep the park service informed about poaching activity and trust that park guards would respond promptly enough to solve any problems before they got out of hand. We ordered more two-way radios so that all of our crew could be in constant contact with one another; that way, someone could be alerted to run to inform the park service when future monkey-poaching activity was discovered. I gave orders that field assistants were to work only in teams, rather than alone as they occasionally did.

While the park service was concerned, it was rather slow to respond. There were various legal reasons why it would be difficult to take action against the man who had killed Nanny. Basically, unless we had multiple witnesses to the actual crime or a videotape of the killing, it would be too difficult to prove in court that he had done what he claimed to have done. Dead bodies and the report of one witness were insufficient evidence, even though the criminal himself had confessed to having committed the crime. The same rules apply to tree poaching: the perpetrator must be caught in the act of cutting; material evidence after the fact, or confessions of having done the illegal logging, are insufficient grounds for punishment.[2] So although the laws theoretically favor the forest flora and fauna, enforcing the laws is virtually impossible. The park service does have the authority to take guns away from people who are in protected areas, and they do this fairly often, but then those poachers are still at liberty to kill animals with rocks.

It soon became evident that this was basically an insoluble problem. We regularly encountered poachers in the forest. Most of the time they were hunting deer or peccaries, not monkeys, and they would quickly vanish into the trees as soon as they spotted us. But what did they do when they encountered monkeys in our absence? All too often, we would return from our monthly five-day vacation to find that a monkey seen in perfect health a few days before was now missing. We followed the monkeys from dawn to dusk for twenty-five days each month, but even the most dedicated primatologists need a break from time to time, and the park guards were understandably reluctant to take our places during our time away. Incredulous, they applauded our efforts as we crashed up and down cliffs and through rivers and wasp-infested thornbrush twelve hours a day, but they preferred to stick to the trails. Fair enough. Unfortunately, the poachers were happy to go wherever the game was; in their own way, they loved wallowing in nature as much as we did.

This was the irony of our predicament—both we and the poachers loved the forest. We loved it in completely different ways, however. What we had in common was a love of fresh air, a willingness to be cov-

ered in mud and organic matter of various sorts, a high tolerance for insect bites and physical discomfort, and a sense of exhilaration after putting in a day of hard physical labor. And we could plausibly bond to
some extent over stories of close encounters with rattlesnakes or killer
bees, though our accounts never ended with the application of a machete to the snake. At the end of a day in the woods, we primatologists
bragged about how many focal follows we had completed and recounted
the best monkey stories of the day to our companions working in the
other groups. In contrast, the poachers bragged about how many animals
they had killed, or how many trees they had cut down right under the
nose of the park service. (Once we found a spray-painted message on a
tree stump, bragging to the park service how much money the poacher
would make from the felled *pochote* tree. The market value of these trees
is higher than the fine for poaching them, which is part of the problem
in enforcing the law.) Whereas we stood in awe of nature and sought to
understand it better and preserve it, most of the men living near the reserve were interested in learning only what would help them exploit the
natural resources more efficiently. The poachers wanted to know the
habits of the animals they were trying to kill, for example, and what tree
species provided the most useful wood. We wanted to understand the
animals on their own terms. We wanted to know what they thought
about, who their friends were, how they decided what to do next, what
they were communicating about, and so forth. Basically, we regarded
them in much the way that most people regard their fellow humans. The
idea of hurting one of them is profoundly repugnant to us.

Whenever I talked to someone whom I knew to be a poacher, I made
an effort to explain that the monkeys are sentient individuals with social
bonds and emotions similar to ours, telling them numerous tales from
the lives of animals I had known for many years. I hoped that this would
establish some sense of empathy in them and deter them from killing
monkeys in the future. More often, however, my stories evoked conde-
scending or disbelieving smiles and a barrage of tales about what monkeys *really* do (rape human women, steal watermelons, tie stolen ears of
corn together while another monkey stands guard). It was clear that
most did not believe our stories about monkey social life any more than
they would believe it if we told them that rocks had social relationships. The fact that we had spent literally thousands of hours systemati-

cally and scientifically documenting the monkeys' social interactions was immaterial. Outlandishly dressed gringo women who spoke imperfect Spanish and spent all day chasing wild animals through the forest, rather than staying home and doing housework like proper women, simply had no credibility.

Even some of our most trusted, respectful, and well-meaning friends in the community occasionally capture endangered forest wildlife to keep as pets, much to our dismay. When we ask them about this, they say that the animal's mother had abandoned it. Apparently this sounds like a plausible story to people who do not study animal behavior. Our patient explanations of how mothers always return to retrieve their offspring once humans stop lurking near the nest, or once they are done with their brief foraging mission, are firmly negated by these pet owners, who continue to insist that they are just doing a good deed by caring for these "abandoned" animals.

Many owners of pet parrots and monkeys quite clearly love their pets and do their best to take good care of them, at least when the pets are young, cute, and temperamentally more manageable. (Most owners of pet monkeys either kill them or turn them loose to die when they reach adulthood and become destructive.) But it is painfully obvious to someone familiar with the animals' behavior in the wild that life in captivity is a pathetic substitute for life in the forest, where they can eat what they are designed to eat (instead of rice and beans) and choose their own social companions. Eventually we came to the conclusion that it was too hard to change adults' minds about keeping such pets, and we resolved that the only way to make an impact was to work with children, whose minds are more open to new ideas and values.

Although poaching and the capturing of young animals for pets are harmful to the region's fauna, fire poses possibly the greatest threat to the environment of the tropical dry forest. *Jaragua* and other African savanna grasses were introduced in Guanacaste province, where Lomas Barbudal is located, in the 1920s for cattle pasture. The young stems and leaves are highly nutritious, but the mature plant is very fibrous, low in nutrition, and not preferred by cattle. Because of this, the mature grass (which is exceedingly dry and flammable during the dry season) is burned off annually by ranchers to make way for new pasture. *Jaragua* grows to two meters high, and when it burns, it produces a much hotter fire than native vegetation does; thus when *jaragua* begins to invade forests, the resulting fires do considerable damage to the undergrowth as well as to the root systems and trunks of thin-barked trees.

A year after a major fire, following the rainy season, which will have rejuvenated the undergrowth, a casual observer may be fooled into thinking that little or no damage has been done. After all, most of the trees will still be standing after a single fire. However, each successive fire damages the root systems of the trees further, resulting in a gradual change in the species composition. Fire favors the growth of grasses and other undergrowth but inhibits the growth of many native forest species, reducing biodiversity and encouraging the propagation of non-native grasses. The park service has rented out the forest to cattle ranchers to use as "pasture" land, arguing that grazing will reduce the amount of undergrowth and thereby minimize the risk of fire. But in the dry season, cattle tend to prefer to forage on native species rather than the invasive grasses, and the species composition of forest patches that have housed cattle becomes dramatically different and less diverse than forest that has not been foraged by cattle.[3,4]

We regularly encounter both cattle and fire in the forest when we are collecting data. The areas frequented by cattle are notable in their species composition, consisting of a small number of very thorny, tough-stemmed plant species by the end of the dry season. Every year we discover fires while collecting data. Sometimes these fires have been started by poachers trying to flush out deer. Other times, the fires have escaped from nearby ranches and farms (especially sugar cane farms), where the managers of the land did not make adequate firebreaks, or chose to burn garbage or pasture on a windy day. It is common for up to 50 percent of the forest to burn in a single dry season, and I cannot recall a single year in which no fires affected the areas used by the capuchins.

In the early years of our project, news of fires used to panic me completely, because I did not know whether the monkeys would know how to escape. I was tremendously curious about how they would react to fire, so I attempted to stay with them during forest fires, at least so long as I felt I was in no great personal danger. I have done focal follows while flaming trees were crashing down nearby, and sometimes my field assistants and I have had to abandon the monkeys because we could not see them through the smoke. Oddly, the monkeys do not react to fire and smoke at all; they will take long siestas on the edge of a burning field, with ash and smoke billowing around them, behaving as if nothing is the matter, while we humans cough and cover our faces with bandanas and eventually give up and leave because we can't breathe or our boot soles are melting. To this day, I do not understand how the monkeys manage to survive the really large forest fires.

Of course, the fact that the monkeys manage to escape death in the flames does not mean that their populations are unaffected by fires. Fire has a largely negative effect on biodiversity, and it is harmful to many of the tree species from which the monkeys obtain fruit. At the same time, secondary growth and rotting wood, two consequences of fire, seem to be favorable habitat for many insect species preferred by the capuchins. Overall, however, it seems that the monkeys tend to avoid parts of the forest that have recently burned and to concentrate their foraging time on areas of riverine forest, which is the least vulnerable to burning.

If a fire has not spread very far and it is not a windy day, we use our machetes to cut firebreaks around the affected area so that the flames cannot spread further. But often we must seek help from the community or the park guards. For many years, the nearest park service station was many kilometers from our study site, and it was far easier and faster to get help from the local farming community or from Hacienda Pelon, the neighboring ranch. We would drive into the community, pick up men in their fields, drive them to the site of the fire, and ferry shipments of water to them as they worked. Often we worked alongside them, but they were so much quicker with a machete that we felt fairly useless in comparison. Also, fire fighting is very much a "male bonding activity," and most often our field assistants have been women. Many of the men did not seem to feel comfortable having women help in the fire fighting, and they preferred to send us off to fetch water so they could make bawdy jokes while they worked, without having to worry about what we thought of their humor. Although it is always depressing to see the forest burn, it is cheering to see so many men banding together to solve a community problem that doesn't directly affect them or their own property. However, when fires are not detected until nightfall, or when it is a windy day, they can get severely out of hand, to the extent that a bunch of men with machetes are powerless to put them out.

It isn't so hard to convince most Costa Ricans that forest fires are bad news. Everyone can appreciate the fact that smoke is unpleasant, that burned forest isn't as aesthetically pleasing as living forest, and that charred root systems are ineffective at preventing erosion. There are clear-cut, practical advantages to just about everyone in keeping fires under control. But there will always be a few people who are pyromaniacs, or who are simply too lazy and irresponsible to properly manage their land by following the park service–prescribed burning procedures.

It is far more difficult to teach people to appreciate forest fauna than it is to teach them to follow proper land-management procedures, because the fauna offer no obvious advantage to them personally. Unlike North American children, who make frequent trips to zoos and read children's stories full of anthropomorphic animals who act (implausibly) just like humans, Tico children (with rare exception) are not taught to value animals as individual, sentient beings. Animals are just another resource to be exploited. Most of the animals they encounter are domestic animals that are being raised for consumption.

Virtually all of the boys in the town where we live make a habit of throwing rocks at any wild animal they see, and unfortunately, when schools make field trips to the forest, the boys continue to exhibit this behavior in there. On many an occasion we have been forced to leap out of the shrubbery to scream at a child (or even an adult) who is poised, rock in hand, ready to kill one of our habituated monkeys just for fun. It is hard to imagine how they would gain any machismo points with their buddies by striking dead a monkey who is calmly sitting on the ground a meter away from them, but the fact that this isn't really a challenge does not seem to deter them.

In a sad coincidence, Nobu, the son of Nanny, was killed by a tourist kid with a rock on the two-year anniversary of his mother's death at the hands of the rock-wielding tree poacher. This was one of the most frustrating events in the history of the project. Nobu had overcome so many incredible challenges as an orphaned eight-month-old juvenile that he had won the hearts of all the project members and become the "poster child" of the monkey project. Nobu was only half the size of his age-mates, and he had lost half his tail fourteen months after his mother's

Abby and her orphaned grandson Nobu walk across the savanna together.
Photo: Whitney Meno.

death. He had also survived capture by Rambo's group and Pelon group in intergroup encounters. It may be that his stunted growth made him as cute to the other monkeys as he was to humans. The females in the group permitted him to nurse far longer than his agemates, and he was also very popular as a grooming and play partner and had no trouble hitching rides from other monkeys.

We were not with Abby's group the day Nobu was killed. Emily Kennedy and Heidi found his body on the "playground" where Abby's and Rambo's groups frequently stop for a siesta and a wild romp on the ground, right by the tourist trail. Rambo's group found Nobu's body and began to inspect it and drag it about. Emily quickly collected it and brought it back to town for an autopsy. The local veterinarian, Dr. Alvaro Rojas Madrigal, kindly provided his services free of charge and confirmed that the death was caused by a sharp blow to the back of the head resulting in a cranial lesion and a seven-centimeter fissure. The body was in perfect condition, aside from this injury (and the fact that it was rapidly beginning to rot).

When I presented Nobu's death certificate to the park guards, they seemed uncomfortable with the idea that this was a human-caused death and suggested that he had just fallen out of a tree. However, capuchins, like cats, always land on their feet when they fall. And Nobu had weighed a mere 1.25 kilograms, which isn't really heavy enough that he could have fatally injured himself in a fall. Monkeys fall or leap from trees all the time during play and fighting and never sustain serious injuries in this way: they instantly charge right up the tree to rejoin the fray. Upon further investigation, we discovered that a boy had been seen throwing rocks on the tourist trail that day, but that no one had followed him to see whether he encountered any wildlife.

Sadly, many visitors to the forest cannot be trusted to behave themselves. Even though the park guards, when they greet tourists and point out the trailhead, ask them to respect the wildlife and refrain from harassing animals, many people flagrantly disobey instructions. To be really certain that tourists follow the rules, it is necessary to send guides with them. This is, of course, far more expensive than simply posting someone in a guard house. But it may be the only way to protect the wildlife that the tourists are coming to see. It is also important that tour guides have enough backbone to chastise their clients when they begin to harass the animals. It is difficult for guides to do this, as it no doubt reduces the size of the tips they receive. All too often I have watched guides laughingly encourage tourists as they goad the howler monkeys into howling by shaking trees, howling at them, and doing Tarzan imitations.

As recounted in Chapter 2, unhabituated capuchins flee from humans. This means that habituation, though a necessary step toward our learning more about them and raising people's awareness of the value of wild monkeys, may increase the capuchins' risk of becoming victims of poaching. The presence of biologists can be the best possible sort of protection for wildlife, but when funding runs out and studies end, we always worry that our study populations will be especially vulnerable to poachers because they have gotten used to us. At Lomas Barbudal, our capuchins are quite skilled at distinguishing us from the local *campesinos,* and they continue to be wary around the latter—but they are not as wary as they were prior to habituation. Every few years we conduct a census of the broader population, in an attempt to assess the effect of our presence on the monkeys. The census data show that the study groups grow faster than the groups that are not studied. It is hard to know whether this is attributable to the deterrent effects of our presence on human poachers or on nonhuman predators. At any rate, the results seem to indicate that we are a net benefit, rather than a net cost, on the monkey population in general.

When tourists are at large in the reserve, our research team members stop collecting focal-follow data and assume the role of wildlife protectors, while continuing the group scans and ad lib data collection. Usually we stay off the trail, in the undergrowth with the animals, and wait to see what will happen before we intervene. Surprisingly, large numbers of talking tourists will walk right through the middle of a group of foraging monkeys without even noticing that they (or we) are present. Or if they do see the monkeys, they say, "Look a monkey! Another one! Another one!" Then they forge on ahead after observing them for a total of about fifteen seconds. Having spent literally thousands of hours watching monkeys, it blows our minds to think that someone would want to spend no more than fifteen seconds watching these fabulous animals when it is possibly their only chance to do so. The best tourists are typically European adult couples who have taken care in organizing their own trip to Costa Rica, and are traveling without a group or a guide. They watch respectfully and stand on the edge of the monkey group for several minutes, taking everything in. When we see that someone has genuine interest, we approach them, briefly explain our work, and offer to answer their questions.

The worst tourists tend to be boys or young men with large local groups, who mainly come to the reserve to swim rather than to learn about the forest. Because they have never traveled to other places, they are unaware that their local wildlife is truly special, and that many of the local species are highly endangered. They perceive them as commonplace and entirely expendable. When they see monkeys, birds, or squirrels, they automatically reach for a rock and let it fly.

These teenage boys are completely baffled when the researchers (typically attractive women in their twenties clad in army pants, tank tops, and snake leggings) leap out of the undergrowth and start berating them. They listen wide-eyed, backing away, as the women heatedly inform them that they know these monkeys on a personal basis. Typically, one field assistant will radio to other researchers that rock-throwing tourists are in the reserve, and someone will be dispatched to notify the park service while others converge on the scene or move to a point further up the trail in front of the next monkey group, to intercept the visitors in case lecture number one is ineffective.

In these instances, our strategy is to talk the visitors' ears off, explaining how important the monkeys are to people from all over the world, and pointing out the parallels to human behavior, in an attempt to make them realize that these animals are worthy of better treatment than being used as a target. We tell them the entire life history of the monkey they almost killed, and point out all of his relatives in the group, as well as his closest friends. We tell the story of Nobu: how he grieved by his mother's dead body, how he was depressed for months, how his relatives adopted him but could not provide him with enough food to prevent his stunted growth. We tell them how many letters of condolence we receive from all over the world when a popular monkey such as Nobu dies.

By this point, the rock throwers are either expressing a bit of true remorse or are convinced that they are surrounded by lunatics. If they seem to doubt our sanity, we try to get one of the Tico members of the project to explain the situation. This is often more effective, since the Ticos can relate better to local tourists, having had conversion experiences themselves with regard to their attitude toward the local fauna.

At some point in the conversation, after having outlined all of the disadvantages to killing these animals, I ask the young men what they have to gain by killing a monkey, and whether they think that those benefits could really compensate for the damage done. Usually they can't think of any benefit to killing a monkey—throwing rocks is just something you do when you see a moving animal. Once again we point out how these animals are highly beloved, not only by their groupmates but also by the human researchers who have come to know them. We mention that monkeys are a big tourist attraction, and that if they are killed off, this will damage Costa Rica's reputation internationally. We explain that, because monkeys give birth to single infants, only once every two years, and that the offspring take six to ten years to mature, the monkey populations recover very slowly from losses compared with faster-breeding mammals such as rabbits or deer. We also tell them that the monkeys at Lomas Barbudal, in particular, are part of a long-term re-search project that will be damaged by poaching. Basically, we say, any-one with pride in the principles that Costa Rica stands for internation-ally should refrain from wantonly damaging the local wildlife.

Sometimes we seem to get through, and the rock throwers apologize and promise to amend their behavior. Other times they retreat in con-fused astonishment from our "lunacy," or (if they outnumber us) they jeer at us. In the worst cases, staff members have been physically threat-ened, though this has happened only twice in my memory, when Ticos from our project have requested that other Ticos modify their behavior and the situation has seemed to turn into an issue of defending one's honor. In one of these instances, Alex ran across a man who was scream-ing and throwing things at the monkeys by the side of the road and pa-tiently asked him to please stop, since he was in a protected area. The man, who was driving a government vehicle (not from the park service, but from some development-related branch of the government), was completely outraged at being told what to do, and he wanted to fight Alex.

Basically, it isn't realistic to think that a short encounter with a few eccentric foreign scientists (or even some converted locals) can undo years of socialization by trusted relatives and friends. What is needed to change Costa Ricans' attitude toward wild animals is persistent environ-mental education during children's early development. If the message is to reach adults, the most plausible way to get it there is for children to transmit their environmentally friendly attitude to their parents.

When Joe and I first arrived in Bagaces, it was home to a superb environ-mental-education program run by UC Berkeley entomologist Gordon Frankie and his wife, Jutta, the couple responsible for getting Lomas Barbudal established as a biological reserve in the 1980s. Gordon and Jutta have made a career of working to preserve the dry forests of Guanacaste, the most northwestern province of Costa Rica. They ran a children's conservation club that met weekly, and since it was practically the only institutionalized form of fun for the children of Bagaces, it be-came immensely popular. The Frankies organized frequent field trips and regularly brought in biologists and other volunteers to teach the children. During my first years of fieldwork, from 1991 to 1993, Joe and I used to volunteer at the center, teach English classes and giving occasional talks about the monkeys to the children's nature club. The youth group was also active in community service and helped to clean up garbage and clear trails in the reserve, among other things.

Years after the Frankies retired and ended their environmental-education program, I still encountered people who were in their club as children and spoke fondly of their time with the group. These adults have notably different attitudes toward nature, compared with other locals. They are quick to volunteer to help maintain the forest reserve or to do environmental cleanup work. Many of them are part of the local volunteer *bomberos* (firefighters) squad, for instance. This group is led by a former high school custodian, and they were initially trained in modern firefighting techniques by Gordon Frankie in 1986. The *bomberos* squad has saved the day many times when the park service had insufficient manpower to extinguish a fire, or when fire broke out in an area outside the official jurisdiction of the park service. The Frankies eventually retired, and their environmental-education program folded for lack of resources and volunteers to continue the work.

With our dawn-to-dusk research schedule and short vacations, we are not in a great position to launch an education program of the same scope that the Frankies managed, but we do our best. During the monthly five-day vacations, as well as on some of my data-transcription days, members of the research team and I often visit the elementary schools neighboring the reserve. The children attending these schools are the people who will ultimately have the greatest impact on the fate of the Lomas monkey population. It is our hope that if we can persuade them that monkeys are worth preserving, they will also persuade their parents and other family members not to kill monkeys and other wildlife.

Our visits to the schools are generally met with great enthusiasm. We bring laptops and show the children videos of the monkeys in the wild. They are enthralled by the sight of monkeys socializing in a relaxed manner. Antipredator behavior is usually the only monkey behavior they see, and so it is commonly believed that monkeys are generally *"bravo"* (mean or aggressive). They enjoy trying to decipher the monkeys' vocalizations, and are fascinated and bemused by the fact that our jobs consist of following monkeys all day. Some ask what kind of education they would need to study monkeys as well. Our accounts of the monkeys' lives are met with a mixture of interest and incredulity, and we have to repeatedly emphasize how we gain this knowledge that seems so at odds with what they have always been told about animals.

Our aim is to create a more personal connection between the students and the monkeys that are their neighbors. To help accomplish this, we also recently started an adopt-a-monkey program, in which each of four elementary schools near Lomas agrees to "sponsor" a baby monkey. We do not request that students donate money but merely that they pledge to protect the monkeys and do their best to promote monkey welfare among their friends and relatives. We provide each school with a poster displaying pictures of its adopted monkey and that monkey's close relatives, and including a short biography of the monkey. Every six months we return to the school with an update on the monkey's life, so that the children can appreciate the kinds of challenges a monkey faces as it matures.

As the program has developed, we have added more layers. For example, we now give each classroom a "monkey ambassador book" that contains copies of the updates and photos of the school's adopted monkey. These books also contain text sections (in Spanish) relating general facts about capuchins and answering commonly asked questions. We urge the students to borrow these books and take them home to show to their families, so they can serve as ambassadors between the monkeys and the community and spread their knowledge more broadly. The books also have spaces in which students and their families can write their own questions about the monkeys for us to answer. The students seem very excited about helping the monkeys, and also pleased that they know things about the local wildlife that their parents did not know. All of them sign a pledge to support the wildlife. We promise to return each year to celebrate their adopted monkeys' birthday with them, provided that the monkey survives the year.

In another part of the monkey ambassador books, we ask the children to write descriptions and make drawings of monkeys they see in the wild, in case they find some of our emigrant males. We also include space in which the children can report situations they have seen that they think might be dangerous to the monkeys. We give the teachers and school directors our contact information so we can be reached in case of a crisis (if someone knows of monkeys being captured for pets, for example, or if a monkey has been hit by a car). The children from the San Ramón school immediately reported a monkey that had been electrocuted near their school. We went to investigate and found that, indeed, there was a monkey literally fried on the wires, in a reclining position. We were afraid that his groupmates would climb up to investigate his odd behavior and suffer a similar fate, so we called the ICE, the electricity company, and workers instantly came out to remove the body.

When electrical cables had first been installed in the northern end of Lomas reserve, we had petitioned the power company to insulate the wires, showing them photographs of monkeys from our population that had been electrocuted on wires at a nearby ranch. The ICE was extremely accommodating and came out with us to see where the monkeys ranged. At great expense, the company insulated all of the cable that ran through forested areas. After the electrocution incident near San Ramón school, the company said it would rewire that section as well. Not only did the ICE insulate the wires, but it also worked with us to develop plans for building wildlife bridges across roads, to reduce the problem of monkeys and other terrestrial fauna being hit by cars. This plan has still not been fully implemented, but we hope that it will eventually be put into action.

The tricky thing about conservation is that, even if teams of highly dedicated people are pouring time and money into protecting a forest and its fauna, their efforts can be largely undone by just one or two uncaring individuals with a match, a gun, or a slingshot. This is depressing and frustrating. Conservation is a *lucha sin fin*—an endless struggle, requiring constant vigilance and action.


Conservation biology is not a career for the faint of heart. The Frankies, for example, were among the most energetic and optimistic people I have ever known when they started their conservation work in Costa Rica. They did not have children to raise and were able to pour 100 percent of their time and energy into saving Lomas Barbudal for posterity, generously donating vast amounts of their personal resources to help the forest and community. Although their efforts were greatly appreciated by the children of the area, many of the adults viewed them with great suspicion; they seemed convinced that the Frankies had to be making piles of money somehow from their conservation work. After enduring roughly twenty years of vicious local gossip, innumerable thefts at both their personal residence and their education center, and even death threats, they eventually retired, completely exhausted and somewhat embittered by the experience, and retreated to the realm of basic science to do the bee research that had initially sparked their interest in the tropical dry forest.

Despite the hardships the Frankies faced, they did a great deal of good. I doubt that the capuchins of Lomas, and the forest that sustains them, would still exist today, were it not for their extraordinary dedication. I myself would never have heard of Lomas Barbudal if they had not vigorously promoted it as a research site and a tourist attraction, both in Costa Rica and in the United States. And they planted the seeds of environmental awareness in the minds of the next generation of adults. But unless others continue their work, their efforts will eventually be negated.

I am happy to say that many of our former field assistants are now planning their own careers as primate conservationists. They all have their own angles they want to pursue. Some want to focus on environmental education. Others work on land management and environmental policy, or on developing monkey-friendly ecotourism. Still others are concerned about stopping the pet trade and creating more centers for wildlife rehabilitation. They all are highly trained in biology, and they know exactly what they are getting into in terms of local politics and culture. They are not quitters: some have been getting up with me before dawn to chase monkeys thirteen hours a day for years on end. Their passion for the animals and the forest will lead them to fight much harder for the lives of these animals than someone without such personal and local experience.

The Lomas Barbudal "*moneros*"—the monkey people—are exceptionally bright, dedicated, and cooperative people. It is hard to imagine such a talented bunch failing at anything they really set their minds to. So I think the capuchins have a chance. It won't be easy, but we have gotten an early start on the mission of showing the world that these creatures are worth saving. In contrast, species such as the orangutan are in danger of being exterminated before we can gain extensive knowledge about the intricacies of their behavioral biology. I hope that readers of this book will transmit what they have learned to other people, especially children. Even if you never make it to Costa Rica to meet the capuchins, you can make a big impact by educating children about the value of wildlife and fostering appropriate interactions with individuals of other species.

# Epilogue

January 1, 2006     I am on the plane, traveling back to my field site after an emotionally draining month at UCLA. My five-year contract
with the Max Planck Institute is almost at its end. It is time to start
transitioning back from "researcher heaven" to the real world, where
there is intense competition for funding and research becomes an expensive hobby—a guilty pleasure sneaked during evenings and week-
ends after teaching duties are done—rather than a job you get paid for. I
have spent the past few months writing grant proposals, teaching, and
publishing articles like mad, so that I will be prepared to keep the field
site running once the institute's funding ends. In the past fifteen years, I
had never failed to obtain outside funding for my site, although we
worked with very small budgets in the pre–Max Planck days. However,
December brought a string of rejection letters from funding agencies. I
have spent the past few weeks frantically analyzing pilot-study data and
rewriting proposals. Now I am preparing to meet my field crew, with a
bit of trepidation: I have nothing but bad news to tell the talented long-
term staff members about their future employment prospects, or about
the prospects for the monkeys' continued safety.

As soon as I get off of the airplane and smell the familiar Costa Rican scents, my spirits lift and I feel my muscles start to relax in the warm, humid air. The tiny Liberia airport is basically an open-air *"bodega"* (warehouse) in the middle of a big field, and I sail through immigration and customs in just a few minutes and look about eagerly for Alex or Hannah. I don't recognize Alex at first because he has once again dyed his hair, this time blond. "Alex! *Que te pasó? Habia otro accidente con el Born Blonde?"* ("What happened to you? Was there another accident with the Born Blonde?") In the past he has accidentally gotten more Born Blonde hair dye on himself than on the monkeys when we were marking them. Alex laughs his contagious belly laugh and explains that this time it was a fully intentional change of hair color. He gives me a hug, helps me load my bulging, overweight suitcases into the car, and we head off toward Bagaces. As we make the thirty-five-minute drive to the project house, he rattles on and on with news about park service politics, the monkeys, the sorry condition of the roads, and the endless ailments of the project vehicle. The Pan-American Highway is, indeed, a mess. At times Alex startles me by swerving all the way off the road to avoid dangerously deep holes stretching across both lanes.

Alex has become truly passionate about conservation in recent years. Whenever I need someone to help me with environmental education, or to attend a park service meeting that I have to miss, he eagerly volunteers. He has shown such an aptitude for conservation politics—with his energy, passion for the wildlife, sense of humor, and ability to communicate well with all sorts of people—that I have made him the official project representative on various committees. He loves giving talks about the monkeys and has all sorts of excellent ideas for sustainable eco-tourism. As we drive, he earnestly explains that he may want to cut back his monkey-watching hours to half-time, so that he can pursue some of his other interests, perhaps starting a *vivero*, a tree nursery for reforestation, or devoting more time to conservation politics. While he loves the monkeys as much as ever, he has been watching them from dawn to dusk for more than five years now and feels he needs a change of pace—but one that will continue to benefit the monkeys.

As we pull up to the gate, Alex cautions me to be ready for a shock: there are many new adopted animals, and also quite a few visitors at the house. "*Es un poco loco ahora,*" he explains ("It's a little crazy now").

Juan Carlos strolls out to the gate to greet me. "*Mucho gente!*" he says, echoing Alex's warning about the crowd of people inside, and he gestures toward the house, shrugging his shoulders and grinning. Juanca loves peace and quiet, just as I do, and he often retreats to the hammock outside when the evenings get too chaotic. But these are people I am eager to see: seven former field assistants have saved their money to fly back to Lomas for a New Year's reunion and to spend a few days with the monkeys. I greet the dogs, push my way through a throng of unfamiliar cats milling about on the porch, and enter the house. More than a dozen people are packed like sardines around the dining table, and five others are standing nearby, all eating their evening meal and engrossed in conversation. I slip quietly into the house and pause for a few seconds to orient myself before joining the conversation.

As I stand by the door, bewildered by the mob in front of me, I am spotted by Brent Pav and Whitney Meno, two field assistants from 2002, and they stand up to greet me. Brent has just returned from Africa, where he was studying hyenas and blue monkeys, and Whitney has also come from Africa, from the Congo, where she was studying gorillas. I work my way through the group, catching up on everyone's news: most of the former assistants have gone on to do fascinating fieldwork on animals living in such exotic places as Ecuador, Panama, Kenya, Botswana, Uganda, and Thailand. Whereas some went directly into further fieldwork, others enrolled in graduate school immediately after leaving our capuchin project. Some, like Gayle Dower and Tom Lord, went into the conservation sector; they are campaigning to preserve wildlife in North America and Great Britain. I am pleased to say that all twenty-three of those students from my project who applied to graduate school were accepted and are currently developing exciting thesis topics—most of them in primatology, but a few are working on other taxa. Some, like Colleen Gault, Nando Campos, and Mackenzie Bergstrom, have caught the capuchin bug and plan to devote their dissertation research, and perhaps their careers, to studying capuchin monkeys.

I am tempted to stay up all night talking, but I am itching to see the monkeys, which means I must be unpacked and field ready by 4:00 A.M. So I start unpacking while continuing our conversation. My suitcase is full of supplies, both essential (repaired equipment, for example) and unessential but highly valued (special ingredients for group dinners and exciting field snacks for the crew). I hand these things over so they can be stowed in the appropriate places. We exchange Christmas gifts, all of which bear the marks of our Lomas capuchin obsession. Then I head to my room, where Hannah and Alex fill me in on the latest monkey news while I get my gear ready, throw a sheet over a foam pad on the floor, and string up a mosquito net. I manage to turn in by 11:00 P.M., but I lie awake for most of the night in eager anticipation of my reunion with the monkeys, listening contentedly to the sounds of the night creatures. Geckos chirp at one another across the room. I can hear termites quietly chewing on the wood of the walls. Roosters crow in the neighbor's yard, and something rather large scampers about in the space between the ceiling and the tin roof—a cat chasing a mouse, perhaps. Home, sweet home. The stress of academia begins to evaporate as I settle back into my Costa Rican life.

The next days are so packed with activity that I hardly have time to worry about the funding issues that were causing me so much anxiety in the preceding months. I need to catch up on the monkeys' lives, and also discuss everyone's future plans with them. It is exciting to hear what past field assistants want to do and to help them develop their thesis topics, even when their plans involve work at other sites. Colleen wants to come back to Lomas to study the physiological underpinnings of bond testing in females, focusing primarily on the eyeball-poking clique of monkeys in Pelon group. Hannah is writing up the infanticide data and making her plans for graduate school, where she will probably focus on conservation. After spending five and a half years working on our project, she has decided it is time to step down from the field manager position and move on, so she can get the training that will help her preserve the habitats she has become so fond of during her time here.

Wiebke Lammers, who will take over as field manager, is developing an exciting plan for a thesis on demographic influences on individuals' social strategies. She will use the long-term database to see how sudden changes in group size (as a consequence of migration or group fission) alter animals' ranks, personalities, foraging success, social dynamics, and reproductive success. We are also starting up collaborations with Tobias Deschner, of the Max Planck Institute, to measure testosterone and cortisol levels in our male subjects as they migrate from group to group, forming new relationships and getting into fights. And Juan Carlos, in collaboration with some chemists at Costa Rica's Universidad Nacional, is avidly pursuing his interest in plant-animal interactions by studying the chemical composition of *Jacquinia* fruits and how it affects animals' decisions to eat them or rub them on their fur.

All of these new projects require some adjustment in data-collection protocols and the creation of makeshift laboratories, so every minute of the day and much of the night is consumed by data collection and discussions with staff. It is a delight to be part of such a hard-working, enthusiastic, and collaborative group of people, and I spend most of each day in a contented glow, except for those moments when staff members ask me about the future of the project.

It is hard for these young scientists, who have known the project only during the Max Planck–funding phase, to comprehend the harsh realities of scientific-research funding in the United States. Over the past seventeen years, the Lomas Barbudal Monkey Project has accumulated the richest continuous behavioral database for any wild primate population, and we even know the genetic relationships of virtually all of the animals in our study population. Having such detailed background knowledge on a large number of individuals and social groups enables us to answer difficult questions about life-history strategies, the influences of individual practice and social learning on the development of skills, personality development, and the reproductive consequences of particular social strategies.

Even a six-month gap in funding wreaks havoc with a project such as this, because it means that valuable staff members must seek jobs elsewhere, and critical information about births, deaths, migrations, and social transmission of traits is lost forever. It is impossible to stop a behavioral study and pick it up later without seriously damaging the quality of the data, particularly if reproductive success is a variable of interest. Not only are important observations missed but also habituated animals are often killed by poachers or caught for the pet trade when their observers are not around to protect them.

Despite such consequences, the fact is there is simply not enough money to go around to keep long-term projects going and also fund projects by promising new researchers. And even though long-term studies clearly provide higher-quality data on a wider range of interesting topics, compared with brand-new projects, which often fail owing to the challenges of habituating animals and obtaining support from the community, long-term studies often flounder for lack of financial support. Many funding agencies think that the money should be spread around rather than focused on a few senior researchers. This policy is understandable: it brings in new blood. But in a field like primatology, it means that many projects are killed off just as they reach their prime and are finally able to yield the sorts of data that the profession most desperately needs to make intellectual advances.

When funding proposals fail or grants run dry, private universities sometimes provide short-term funding to serve as a safety net, to keep field projects afloat while researchers scramble for more grant money. But public universities such as my own are not so well funded and typically do not provide this type of support. Many a talented field primatologist has been forced to close a site and retire to the armchair because of a lack of funding or institutional support.


A few weeks into the 2006 field season I receive an e-mail from the National Geographic Society, telling us that it will provide a year-long grant sufficient for me hire Juanca and Alex half-time and also cover living expenses for Wiebke and a volunteer or two. Relief washes over me, and I dash off a note of profuse thanks for this desperately needed support. For one more year, at least, we will be able to keep the project running and protect the monkeys, and in the meantime we can scramble for other sources of money. In the current U.S. funding climate, it is unlikely that I will ever be able to obtain salary support, either for myself or for my hard-working and loyal staff of volunteers. But we are addicts now, far too attached to the monkeys and to the research to give up on the project until our personal resources have run completely dry. It would be impossible to turn our backs on the monkeys, now that we have made them so much a part of our lives.

# Afterword to the 2011 Printing

Although only three years have passed since the original publication of this book, much has happened in the lives of the monkeys. Diablita is alpha female of Abby's group once again, having defeated her sister Vandal. Alpha male Tranquilo was killed, apparently by his own brothers, and Duende became alpha male in 2007. To our surprise, Duende sometimes commits infanticide, even though the infants are presumably his brother's offspring. Rambo's group has undergone more dramatic changes. Pablo, their alpha male of eighteen years, was overthrown by Moth. Pablo produced (minimally) twenty-five offspring, eighty-three grand-offspring, and nine great-grandoffspring (quite an achievement for a mammal!), and four of his sons have enjoyed long stints as alpha males in various groups. When he died in 2008, eight of his male descendants left the group to form an all-male group and are unsuccessfully attempting to join a group with females. So whereas Rambo's group was composed almost entirely of descendants of Pablo for many years, it now includes three pairs of adult male co-migrants (the brothers Cassie and Yasuni, Took and Bentley, and a pair of young males). Moth, the alpha male, appears to be the only group member with no adult close kin ties.

Fonz, the alpha male of Pelon group, died after seventeen years as alpha and, like Pablo, ended his life with an impressive reproductive score-card. His death resulted in a group fission.

We are pleased to be able to share the monkeys with the public via film as well as books, now that we have entered a long-term collaboration with our former volunteer, Keith Heyward, who has taken on the task of making documentaries about the monkeys. To see recent photos and video of Abby's group (the "next generation"), visit http://www.prehensileproductions.com. A Spanish-language version of the eleven-minute documentary can be found at http://vimeo.com/13409519.

Finally, if you feel you have learned something of value from this book, we ask for your help. Long-term, individual-based field studies such as this one are valuable because many of the most important questions in evolutionary biology—questions about life history strategies or changes in population structure—require data sets extending over the lifespan of multiple generations and complete demographic data (including accurate data on genetic relatedness and lifetime reproductive success) for large numbers of individuals. However, the primary source of funding for field studies is short-term grants, which makes continuous study difficult. Gaps in observation or in the demographic monitoring are devastating to long-term studies and greatly reduce their value.

As the state of the economy declines, the Lomas Barbudal Monkey Project depends increasingly on volunteer labor and external funding to survive. We encourage those of you who would like to help the project, either with a financial contribution or by volunteering your time, to visit the project website: http://www.sscnet.ucla.edu/anthro/faculty/sperry.

# Cast of Characters

*This is not a complete demographic summary but merely a list of those groups and monkeys mentioned by name in the book (except for a few minor individuals). Important characters are in boldface.*

## Groups

*Abby's group:* The original study group at Lomas Barbudal, habituated in 1990.

*Chingo's group:* An unhabituated group on the edge of Lomas Barbudal, overlapping with Rambo's, Solo's, and Abby's ranges.

*CP group:* A group of monkeys at Santa Rosa National Park in a long-term study conducted by Linda Fedigan and her students.

*El Salto group:* A group living adjacent to Pelon group (Fonz's group).

*Flakes group:* A fission product of Abby's group, formed in late 2003. This group resides in Hacienda Pelon, overlapping in home range with Pelon group.

*Muskateers group:* A fission product of Rambo's group, formed in 2004. Its home range overlaps primarily with those of Pelon group, Rambo's group, and Splinter group.

*Newman's group:* A small group adjacent to Pelon group and Muskateers group.

*Pelon group:* A large group residing in Hacienda Pelon (adjacent to Lomas Barbudal Biological Reserve). It was first habituated in 2002 but probably had the same alpha male, Fonz, for a decade prior to habituation.

*Rambo's group:* A group residing in the Río Cabuyo valley. Its home range overlaps extensively with Abby's group. Pablo has been alpha male of this group since long before the group was fully habituated in 1996.

*Salome's group (also known as Solo's group):* Probably a fission product of Chingo's group, to which several males from Abby's group migrated.

*Splinter group:* A fission product of Rambo's group that seceded in late 1999.

*Vaqueros group:* A group that invaded the Quebrada Congo valley and fought often with Rambo's group in 1997.


## Individuals

*Entry format: sex, birth group, and birth year, followed by other comments (A = Abby's group, F = Flakes, M = Muskateers, P = Pelon, R = Rambo's).*

**Abby:** Elderly female in Abby's group at the time it was habituated in 1990. Alpha female from 1990–1998. Dies in 2003.

**Alamasy:** Male, P~1994. Moves briefly to Flakes group in 2003–2004, then returns to Pelon group.

*Albert:* Sex unknown, A1997. Cujo's twin, born to Abby. Dies in 1997.

*Al Gore:* Male, A1999. Nanny's son. Emigrates in 2003 as part of an all-male group.

**Aramis:** Adult male of unknown age and parentage. Migrates into Rambo's group with Porthos and Athos in 1997. Becomes alpha male of Muskateers group in 2004.

*Athos:* Male of unknown age and parentage. Migrates into Rambo's group with Porthos and Aramis in 1997. Vanishes in 1999.

*Bailey:* Male, A1999. Abby's last son. Dies in 2003.

**Bentley:** Male of unknown age and parentage. Migrates into Abby's group in 2002 as an adult and emigrates in 2003. Occasionally seen with Took and other males in all-male groups thereafter.

*Boones:* Female, A1999. Goes with Flakes group during the fission of Abby's group.

*Broma:* Female, A1991. Vanishes in 1996.

*Butler:* Sex unknown, A1997. Dies soon after birth.

*Calabaza:* Adult female of unknown age, natal to Pelon group.

*Camden:* Female, P2002.

*Capulet:* Female, P2001. Calabaza's daughter. Dies in 2005.

*Cassie:* Male, A2000.

*Celeste:* Female, A1993. Diablita's daughter.

*Champignon:* Adult male of unknown parentage residing in Pelon group.

*Chaos:* Female, R2002. Cupie's daughter. Accompanies her parents in joining Muskateers group during the fission of Rambo's group.

*Chingo:* Alpha male of Chingo's group.

*Cimo:* Male, A1994. Emigrates in 2001 to join Rico, Dos, and Kola as part of an all-male group. Subsequently seen in Newman's group and Solo's group.

*Cookie:* Female, A1999. Vandal's daughter.

*Cricket:* Sex unknown, A2003. Celeste's infant. Infanticide victim.

*Cupie:* Adult female of unknown age born into Rambo's group. Goes with Muskateers group during the fission.

*Cujo:* Sex unknown, A1997. Albert's twin, born to Abby. Dies a few months later.

**Curmudgeon:** Adult male of unknown age and parentage. Alpha male of Abby's group for at least seven years. Deposed in 1992, he stays in the group for four more years until becoming a lone male. Last seen in 1997.

*Dali:* Female, A2001. Diablita's daughter.

*Delilah:* Adult female of unknown age and parentage, natal to Rambo's group. Becomes alpha female of Muskateers group following the fission.

**Diablita:** Female, A~1987. Abby's daughter. Becomes alpha female in 1998, approximately, and remains alpha female until 2004.

**Doble:** Adult male of unknown age and parentage, present in Rambo's group from 1996 to 2003.

*Dos:* Male, A1993. Emigrates in 2001, joining Rico, Cimo, and Kola in an all-male group before settling in Newman's group.

**Duende:** Male, R1996. Migrates to Abby's group in 2003.

*Dusty:* Alpha male of Vaqueros group in 1997.

*Eggnog:* Sex unknown, A2003. Eldritch's offspring. Probably killed by Punto a few months after birth.

*Einstein:* Sex unknown, A2002. Eldritch's offspring. Killed by Jackson four months after birth.

*Eldritch:* Female, A1992. Tattle's daughter. Goes with Flakes when Abby's group fissions.

**Elmo:** Male, R1996. Mezcla's son. Migrates to Abby's group with Duende and Tranquilo in 2003.

**Feo:** Elderly male of unknown age and parentage residing in Rambo's group. Probably resident there since before 1991. Vanishes in 1997.

*Fester:* Sex unknown, A1991. Tattle's infant. Probably eaten by a snake in 1992.

*Fink:* Male, A1997. Tattle's son. Emigrates with several other natal males in 2003.

*Fishy:* Female, A1999. Maní's daughter.

**Fonz:** Male of unknown age and parentage. Long-term alpha male of Pelon group.

**Frodo:** Male, A2003. Diablita's son. Dies in 2003; probable infanticide victim.

**Gandalf:** Male, P~1995. Dauphin's son. Migrates to Abby's group briefly before returning to Pelon group.

**Guapo:** Adult male of unknown age and parentage residing in Abby's group, 1990–2000. Alpha male briefly in 1998 and then for a longer period in 1999 and 2000. Inventor of many games. Vanishes in 2000.

*Heinrich:* Adult male of unknown age and parentage. Migrates from Pelon to Abby's group in 2003; ultimately becomes alpha male of Flakes group.

*Hobbes:* Male, A~1989. Vanishes by 1996.

**Hongo:** Adult male of unknown age and parentage. Migrates into Abby's group in 1996 and stays until 2004. Alpha male for many brief periods in 1998 and 2000 to 2002.

**Ichabod:** Adult male of unknown age and parentage (possibly natal to Rambo's group). Migrates into Abby's group in November 1992 and stays until late 1995/early 1996. Disappears for several months in 1996 and returns in late 1996 as alpha male of Abby's group. Abby's group evicts and then kills him in March 1998.

**Jackson:** Adult male of unknown age and parentage. Seen in Chingo's group prior to immigrating to Abby's group in May 2002. He is joined shortly afterward by Jordan, who may also have originated in Chingo's group. Jackson is alpha male for several short periods during 2002 and 2003. He and Jordan depart from Abby's group for the last time in August 2003. They make a few appearances in Flakes group in the subsequent two months and are seen together a couple more times in the months after that, but have not since invaded a bisexual group, to the best of our knowledge.

*Jade:* Male, A2001. Jester's son. Goes with Flakes group during the fission in 2003.

**Jester:** Female, A1990. Wiggy's daughter. Becomes alpha female of Flakes group after Tattle's death in 2004.

*Jinx:* Sex unknown, A2003. Jester's infant. Probable infanticide victim.

**Jordan:** Male of unknown age and parentage who migrates into Abby's group in 2002. Jackson's sidekick. He comigrates once more with Jackson when Jackson leaves Abby's group for good.

**Kola:** Male, A~1988–89. Wiggy's son. He leaves Abby's group in 1996 to spend time in an all-male group with Quizzler, but returns to Abby's group in 1999 for another two years. He forms another all-male group with Rico, Cimo, and Dos, and eventually is killed by Rambo's group in 2002.

*Lefty:* Young adult male in Vaqueros group.

*Lizano:* Male, A1991. Nanny's son. Vanishes during an intergroup encounter in 1998.

*Lucky:* Young adult male in Vaqueros group.

*Macadamia:* Sex unknown, A2003. Maní's infant. Infanticide victim, most likely killed by Moth.

**Maní:** Female, A1990.

*Marañon:* Male, A2001. Son of Maní. Vanishes in 2005.

*Mead:* Female, A1999. Goes with Flakes group during the fission of Abby's group.

*Mezcla:* Female of unknown age and parentage. Alpha female of Rambo's group.

*Miffin:* High-ranking adult female born in approximately 1992 in Pelon group.

*Minstrel:* Female, A1997. Goes with Flakes during the fission.

*Mischief:* Sex unknown, A1995. Vandal's offspring. Dies in infancy.

*Mooch:* Female, R1999. Daughter of Mezcla and Pablo.

*Moth:* Male, R~1992. Alpha male of Abby's group very briefly in 2003 before returning to Rambo's group.

*Mowgli:* Female, P2003. Miffin's daughter.

*Nanny:* Female, A~1985. Abby's daughter.

*Ned:* Young adult male in Vaqueros group.

*Newman:* Male, R~1991. Pablo's son. Leaves in 2002 to become alpha male of Newman's group, where he is coresident with Cimo and Dos for a while.

*Nobu:* Male, A2001. Nanny's son. Orphaned at age eight months when his mother is killed by a rock-wielding poacher. Killed by a rock-wielding tourist in 2004.

*Nut:* Male, A1996. Maní's son. Emigrates in 2003 to join Salome's group, where he is coresident with Solo, Cimo, and Took.

*Oden:* Young natal male in Pelon group.

*Omni:* Sex unknown, A1993. Abby's offspring. Dies three days after birth, of an umbilical hernia.

*Opie:* Female, A1994. Nanny's daughter.

*Ox:* Male, A2003. Opie's infant. Infanticide victim.

*Pablo:* Adult male of unknown age and parentage. Long-term alpha male of Rambo's group.

*Paul Bunyan:* Adult male of unknown age and parentage. Resident of Abby's group from 1990 to 1995. Alpha male from 1992 to 1995.

*Pobrecito:* Sex unknown, stillborn, A2004. Maní's offspring.

**Porthos:** Adult male of unknown age and parentage. Immigrates into Rambo's group in 1997, along with Aramis and Athos. He goes with Muskateers during the fission of Rambo's group in 2004.

**Power:** Male, P~1993.

**Punto:** Male, P~1991. Alpha male of Abby's group for a short while in 2003.

**Q:** Young male of unknown parentage. Resident in Rambo's group in 1997.

**Quizzler:** Male, A~1987–88. Tattle's son. Alpha male of Abby's group in the summer of 1996, and then joins an all-male group with Kola.

*Rain:* Male, A1999. Eldritch's son.

**Rambo:** Male of unknown age and parentage, born in Rambo's group. Emigrates in 1997.

**Rico:** Male, A1991. Emigrates in 2001 to join Kola, Cimo, and Dos in an all-male group. Killed by a car in 2002, during an intergroup encounter.

*Ripley:* Adult female of unknown age and parentage. Peripheral, subordinate female in Rambo's group.

*Rumor:* Elderly nulliparous female, Pelon group. Probable inventor of eyeball poking.

*Scooby:* Female, R1997.

*Shylock:* Female, P2001.

*Simba:* Male, P1998.

*Simpson:* Female, R1996. Goes with Muskateers group during the fission of Rambo's group in 2004.

*Solo:* Male, A1995 (Tattle's son). Emigrates in 2003 and becomes alpha male of Salome's group, where he is joined by Nut and Cimo.

**Squint:** Adult female of unknown age and parentage, born into Abby's group. Abby's constant companion and ally. Vanishes in 1998.

**Tattle:** Low-ranking adult female of unknown age and parentage in Abby's group. Becomes alpha female of Flakes group in 2003 when the fission occurs. Dies in 2004.

**Thornhill:** Elderly and asymmetrical adult male of unknown parentage who immigrates to Abby's group in late 1998. He has at least two brief stints as alpha male. Vanishes in 2003.

*Till Eulenspiegel:* Male, A1995. Nanny's son. He emigrates in the company of several other natal males in 2003.

**Took:** Wandering adult male of unknown origins who, at different times, is a member of Chingo's group, Abby's group (during parts of 2002–2003), and Salome's group, as well as several all-male bands. Bentley's good friend.

*Toulouse:* Male, A2001. Tattle's son. Goes with Flakes during the fission in 2003.

**Tranquilo:** Male, R~1992. Mezcla's son. Migrates to Abby's group in 2003, where he becomes alpha male.

*Ugali:* Male, R2002. Una's son. Spends some time in Muskateers group before returning to Rambo's group in 2004.

*Una:* Female, R1994. Goes with Muskateers group during the fission.

**Vandal:** Female, A1989. Becomes alpha female of Abby's group in 2004.

*Vishnu:* Female, A2001. Vandal's daughter.

**Vodka:** Sex unknown, A2003. Vandal's infant. Killed soon after birth by Tranquilo.

**Weasley:** Adult male of unknown parentage who migrates into Abby's group in 2001. He is alpha male for some short periods in 2002 and 2003. Vanishes in 2003.

*Wembley:* Male, R2002. Stays in Rambo's group after his mother leaves with Muskateers in 2004.

**Wiggy:** Adult female of unknown age in Abby's group. Dies in 1996.

*Yasuni:* Male, A2001. Opie's son.

*Yoyo:* Male, A1991. Wiggy's son.

# Timeline of Events in Abby's and Rambo's Groups

## Abby's Group

1990: Abby's group is first habituated to human observers. Curmudgeon is alpha male, and the only other adult males are Paul Bunyan and Guapo. Abby is alpha female.

1991: Hand sniffing becomes fashionable among Abby's-group females.

November 1992: Curmudgeon is deposed as alpha male by Paul Bunyan. Ichabod immigrates to Abby's group.

Early 1996: Curmudgeon, Paul, Hobbes, Ichabod, and Kola vanish from Abby's group. Quizzler becomes alpha male, and Hongo immigrates.

Late 1996: Ichabod returns and becomes alpha male. Wiggy dies. Quizzler emigrates to join Kola and an unknown male in an all-male group.

1997: Curmudgeon roams the forest as a solitary male, occasionally joining other migrating males in temporary associations. Squint vanishes, and hand sniffing goes out of style.

1998: Diablita becomes alpha female after a long period of vying with Nanny. Ichabod is killed by the group in March. Guapo and Hongo struggle for the alpha male position, and Guapo eventually wins. Toward the end of the year, Thornhill immigrates and Kola returns to his natal group.

Late 2000: Guapo vanishes, and Hongo becomes alpha male.

Late 2000/Early 2001: Natal males Cimo, Dos, Kola, and Rico emigrate and form an all-male group.

Jan 2001: Weasley immigrates.

Early 2002: There is a rash of visits and immigrations by previously unknown males: Took, Bentley, Igor, Jackson, and Jordan arrive. Hongo, Jackson, Weasley, and Thornhill struggle over the alpha position, and political chaos reigns for the rest of the year. Jackson kills Eldritch's infant.

Early 2002 (all-male group): Rambo's group kills Kola, and Rico is hit by a car during an intergroup encounter with Pelon group. Cimo and Dos join Newman's group.

Early 2003: Abby's group starts to receive frequent visits from various males residing in Rambo's group (mainly Moth and Tranquilo) and Pelon group (mainly Punto, Alamasy, Heinrich). Abby dies.

April 2003: The month starts with a fateful intergroup encounter with Rambo's group, which scatters Abby's group so widely that they never succeed in reuniting completely. Thornhill vanishes, and the Lomas monkey population exhibits a fission-fusion structure throughout the month as males take advantage of the chaos to investigate their migration options. When the dust settles, all seven of the natal males above age four have left the group. One (Solo) becomes alpha male of a small neighboring group, where he is ultimately joined by three male kin. Even the females exhibit a lack of cohesiveness, generally splitting along matrilines.

May 2003: Moth and Tranquilo (natal males from Rambo's group), Heinrich (from Pelon group), and Took (a frequent visitor to many neighboring groups) all visit or immigrate to Abby's group.

June 2003: Moth becomes alpha male of Abby's group and commits infanticide. Jackson returns and defeats him, eventually evicting him from the group. At the end of June, Abby's group is invaded by five Pelon-group males (Punto, Alamasy, Heinrich, Power, and Gandalf). Punto becomes alpha male and also kills babies. Male membership in the group is highly unstable throughout 2003, with twenty-seven different males putting in appearances and up to twelve males present on any given day.

August 2003: A fission process that began with the intergroup encounter in April finally solidifies, and the two lowest-ranking matrilines (Jester's and Tattle's) split off to form Flakes group, with Tattle as alpha female.

September 11, 2003: Punto attacks Vandal's infant Vodka. All of the group's current resident males inexplicably vanish from Abby's group.

September 12–13, 2003: After nearly two years of struggles among all of the neighboring males in the region to be the occupants of Abby's group, three young natal males from Rambo's group (Tranquilo, Elmo, and Duende—all brothers) waltz in when no other males are present and take over the group without a fight. Tranquilo becomes alpha male. He finishes killing the infants in this year's birth cohort. Diablita, the current alpha female, hides on the periphery of the group in a vain attempt to protect her infant, while her long-time ally and beta female, Vandal, becomes acquainted with Tranquilo.

September–October 2003 (Flakes group): Jackson and Jordan are originally the male residents of Flakes group, but they are replaced by Heinrich and Alamasy (originally Pelon-group males) and are occasionally visited by Power, Gandalf, and Punto.

Early 2004: Vandal usurps Diablita as alpha female, with Tranquilo's help. The new males father the next birth cohort, and once again group life settles into a peaceful routine with no male turnover.

February 2004 (Flakes): Tattle dies and Jester becomes alpha female.

March–November 2004 (Flakes): Alamasy leaves to return to Pelon group. Three young previously unknown males (Quixote, Baloo, and Napoleon) independently immigrate to Flakes group.

### Rambo's Group

1996: Habituation completed. Pablo is alpha male (and remains alpha to this day).

1997: Porthos, Aramis, and Athos immigrate. Over the course of the next eighteen months, Feo, Q, Rambo, and Athos disappear.

Late 1999/early 2000: Several females fission off from Rambo's group to form Splinter group. Doble vanishes for a period of nearly a year and then returns.

2002: Pablo's son Newman leaves to become alpha male of a neighboring group.

2003: Moth leaves temporarily to become alpha male of Abby's group, but is evicted and returns. Tranquilo, Duende, and Elmo leave to join Abby's group.

2004: Delilah's matriline breaks off from Rambo's group, along with Porthos and Aramis, to become Muskateers group, and Aramis becomes their alpha male.

Early 2005: Several Rambo's-group males make prolonged visits to Splinter group, and Pablo's son Bart finally leaves to become alpha male of that group.

# Glossary of Behavioral Terms
## in the Capuchin Communicative Repertoire

**Alarm call:** A call produced to announce the presence of a predator. (Different variants are used for different types of predators.)

**Bray:** A call produced by monkeys in a state of extreme agitation, most often during intergroup encounters.

**Burst, or "Let's go" vocalization:** A call emitted in combination with loud trills, to coordinate group movement.

**Coo:** A tonal call produced mainly by infants who are looking specifically for their mothers.

**Dance:** Two monkeys pace back and forth, vocalizing (via grunts, squeaks, or wheezes), pirouetting (spinning), and maintaining eye contact. This sequence generally precedes copulation.

**Duck face:** A pursed-lips facial expression performed during courtship.

**Eyeball poking:** One monkey grasps the hand of a companion and inserts the companion's finger deep into her own eye socket, often up to the first knuckle, usually between the eyeball and the lower lid.

**Food call:**   A call used to signal possession of food (or anticipation of possession of food).

**Food interest:**   One monkey closely observes the hands and face of a foraging monkey from a distance of a few centimeters.

**Gargle:**   A raspy vocalization performed primarily by young infants and their mothers, directed toward adult males (particularly the alpha male).

**Grunt:**   A vocalization performed in a rhythmic exchange by dancing monkeys.

**Hand sniff:**   One monkey inserts his or her fingers into the nostrils (or cups a hand over the nose) of another monkey with whom the monkey has a particularly affiliative relationship.

**Headflag:**   A quick jerk of the head, motioning between an enemy and a prospective ally; used to solicit aid in a confrontation.

**Lost call:**   A loud call performed either by lost monkeys seeking their group or by group members who have noticed the absence of another member.

**Machine-gun vocalization:**   A soft staccato call directed by alloparents toward infants.

**Overlord:**   Two monkeys stack on top of each other to threaten a common enemy, thereby signaling their alliance with each other.

**Peep:**   A tonal call emitted by monkeys who are socializing, especially when requesting grooming.

**Pirouette:**   A monkey spins and flails his or her upper body about while maintaining eye contact with another, during a dance.

**Sex squeak:**   A vocalization emitted while making a duck face and dancing.

**Tooth grinding:**   A monkey grinds or clacks his teeth audibly, most often while staring down an opponent during an intergroup encounter.

**Trill:**   A tonal, staccato sequence of vocalizations used either to coordinate group movement or to signal friendly intentions during social interactions.

**Urine washing:**   Splashing of urine on the soles of the hands, feet, and tail to convey olfactory signals.

**Wheeze:**  A vocalization often performed by dancing monkeys during times of relationship tension.

**Wrinkly face:**  The monkey scrunches up his or her face. This expression often accompanies gargling and lip smacking (when infants do it to adult males) or dances and wheezes.

# Works Cited

## Prologue

1. Fragaszy, D. M., E. Visalberghi, and L. M. Fedigan. *The Complete Capuchin: The Biology of the Genus Cebus*. Cambridge: Cambridge University Press, 2004.
2. Darwin, E. *Zoonomia; or, Laws of Organic Life*. London: J. Johnson, 1794.
3. Romanes, G. J. *Significant Contributions to the History of Psychology 1750–1920*. Washington, DC: University Publications of America, 1883/1977.
4. Belt, T. *The Naturalist in Nicaragua*. Chicago: University of Chicago Press, 1874/1985.
5. Klüver, H. *Behavior Mechanisms in Monkeys*. Chicago: University of Chicago Press, 1933.
6. Klüver, H. "Re-examination of implement-using behavior in a *Cebus* monkey after an interval of three years." *Acta Psychologica* 2 (1937): 347–397.

## 2. The Social Intelligence Debate and the Origins of the Lomas Barbudal Monkey Project

1. Byrne, R. *The Thinking Ape: Evolutionary Origins of Intelligence*. Oxford: Oxford University Press, 1995.
2. Stephan, H., G. Barbon, and H. D. Frahm. "Comparative size of brains and

brain components." In: *Comparative Primate Biology*, ed. H. D. Steklis and J. Erwin, pp. 1–39. New York: Wiley-Liss, 1988.

3. Milton, K. "Distribution patterns of tropical plant foods as an evolutionary stimulus to primate mental development." *American Anthropologist* 83 (1981): 534–548.

4. Parker, S. T., and K. R. Gibson. "A developmental model for the evolution of language and intelligence in early hominids." *Behavioral & Brain Sciences* 2 (1977): 367–408.

5. Jolly, A. "Lemur social behavior and primate intelligence." *Science* 153 (1966): 501–506.

6. Humphrey, N. K. "The social function of intellect." In: *Growing Points in Ethology*, ed. P. Bateson and R. A. Hinde, pp. 303–317. Cambridge: Cambridge University Press, 1976.

7. Byrne, R., and A. Whiten, eds. *Machiavellian Intelligence: Social Expertise and the Evolution of Intellect in Monkeys, Apes and Humans*. Oxford: Oxford University Press, 1988.

8. Cheney, D. L., and R. M. Seyfarth. *How Monkeys See the World*. Chicago: University of Chicago Press, 1990.

9. De Waal, F. *Chimpanzee Politics*, 1st ed. New York: Harper and Row, 1982.

10. Whiten, A., and R. Byrne. "Tactical deception in primates." *Behavioral & Brain Sciences* 11 (1988): 233–244.

11. Dunbar, R. I. "Functional significance of social grooming in primates." *Folia Primatologica* 57 (1991): 121–131.

12. Janzen, D. H., ed. *Costa Rican Natural History*. Chicago: University of Chicago Press, 1983.

13. Perry, S. "Female-female social relationships in wild white-faced capuchin monkeys, *Cebus capucinus*." *American Journal of Primatology* 40 (1996): 167–182.

14. Perry, S. "Male-female social relationships in wild white-faced capuchins (*Cebus capucinus*)." *Behaviour* 134 (1997): 477–510.

15. Perry, S., H. C. Barrett, and J. H. Manson. "White-faced capuchin monkeys exhibit triadic awareness in their choice of allies." *Animal Behaviour* 67 (2004): 165–170.

16. Hare, B., et al. "Do capuchin monkeys, *Cebus apella*, know what conspecifics do and do not see?" *Animal Behaviour* 65 (2003): 131–142.

17. Fujita, K., H. Kuroshima, and T. Masuda. "Do tufted capuchin monkeys (*Cebus apella*) spontaneously deceive opponents? A preliminary analysis of an experimental food-competition contest between monkeys." *Animal Cognition* 5 (2002): 19–25.

18. Mitchell, R. W., and J. R. Anderson. "Pointing, withholding information, and deception in capuchin monkeys (*Cebus apella*)." *Journal of Comparative Psychology* 111 (1997): 351–361.

19. Janson, C. H. "Experimental evidence for spatial memory in foraging wild capuchin monkeys, *Cebus apella*." *Animal Behaviour* 55 (1998): 1229–1243.

20. Ottoni, E. B., and M. Mannu. "Semifree-ranging tufted capuchins (*Cebus apella*) spontaneously use tools to crack open nuts." *International Journal of Primatology* 22 (2001): 347–358.

### 3. The Challenges of Foraging and Self-Medication

1. Perry, S., and L. Rose. "Begging and transfer of coati meat by white-faced capuchin monkeys *Cebus capucinus*." *Primates* 35 (1994): 409–415.

2. Perry, S., and J. C. Ordoñez J. "The effects of food size, rarity, and processing complexity on white-faced capuchins' visual attention to foraging conspecifics." In: *Feeding Ecology in Apes and Other Primates,* ed. G. Hohmann, M. M. Robbins, and C. Boesch, Cambridge Studies in Biological and Evolutionary Anthropology, pp. 203–234. Cambridge: Cambridge University Press, 2006.

3. Beck, B. *Animal Tool Behavior: The Use and Manufacture of Tools by Animals.* New York: Garland Press, 1980.

4. Panger, M. "Object-use in free-ranging white-faced capuchins (*Cebus capucinus*) in Costa Rica." *American Journal of Physical Anthropology* 106 (1998): 311–321.

5. Fragaszy, D. M., E. Visalberghi, and L. M. Fedigan. *The Complete Capuchin.* Cambridge: Cambridge University Press, 2004.

6. Visalberghi, E., and L. Limongelli. "Lack of comprehension of cause-effect relations in tool-using capuchin monkeys (*Cebus apella*)." *Journal of Comparative Psychology* 108 (1994): 15–22.

7. Baker, M. "Fur rubbing: use of medicinal plants by capuchin monkeys (*Cebus capucinus*)." *American Journal of Primatology* 38 (1996): 263–270.

8. Valderrama, X., et al. "Seasonal anointment with millipedes in a wild primate: A chemical defense against insects?" *Journal of Chemical Ecology* 26 (2000): 2781–2790.

9. Verderane, M. P, et al. "True anting in a semifree-ranging group of brown capuchin monkeys (*Cebus apella*)." *International Journal of Primatology* 28 (2007): 47–53.

### 4. Predators, Prey, and Personality

1. Krebs, J. R., and N. B. Davies, eds. *Behavioural Ecology: An Evolutionary Approach.* Oxford: Blackwell, 1991.

2. Sih, A., et al. "Behavioral syndromes: an integrative overview." *Quarterly Review of Biology* 79 (2004): 241–277.

3. Rose, L. M. "Vertebrate predation and food-sharing in *Cebus* and *Pan*." *International Journal of Primatology* 18 (1997): 727–765.

4. De Waal, F. B. M., and M. L. Berger. "Payment for labour in monkeys." *Nature* 404 (2000): 563.

5. Boesch, C., and H. Boesch. "Hunting behavior of wild chimpanzees in the Taï National Park." *American Journal of Physical Anthropology* 78 (1989): 547–573.

6. Perry, S., et al. "White-faced capuchins cooperate to rescue a groupmate from a boa constrictor." *Folia Primatologica* 74 (2003): 109–111.

7. Rose, L. M. "Benefits and costs of resident males to females in white-faced capuchins, *Cebus capucinus*." *American Journal of Primatology* 32 (1994): 235–248.

8. Perry, S., et al. "Traditions in wild white-faced capuchin monkeys." In: *The Biology of Traditions: Models and Evidence*, ed. D. M. Fragaszy and S. Perry, pp. 391–425. Cambridge: Cambridge University Press, 2003.

9. Cheney, D. L., and R. M. Seyfarth. *How Monkeys See the World*. Chicago: University of Chicago Press, 1990.

10. Mineka, S., and M. Cook. "Immunization against the observational conditioning of snake fear in rhesus monkeys." *Journal of Abnormal Psychology* 95 (1986): 307–318.

11. Cook, M., and S. Mineka. "Observational conditioning of fear to fear-relevant versus fear-irrelevant stimuli in rhesus monkeys." *Journal of Abnormal Psychology* 98 (1989): 448–459.

12. Rose, L. M., et al. "Interspecific interactions between white-faced capuchins (*Cebus capucinus*) and other species: Preliminary data from three Costa Rican sites." *International Journal of Primatology* 24 (2003): 759–796.

13. De Resende, B. D., et al. "Interaction between capuchins and coatis: Nonagonistic behaviors and lack of predation." *International Journal of Primatology* 25 (2004): 1213–1224.

14. Janson, C. H. "Capuchin counterpoint: Divergent mating and feeding habits distinguish two closely related monkey species of the Peruvian forest." *Natural History* 95 (1986): 45–52.

### 5. Capuchin Communication

1. Lancaster, J. *Primate Behavior and the Emergence of Human Culture*. New York: Holt, Rinehart and Winston, 1975.

2. Dawkins, R., and J. R. Krebs. "Animal signals: Information or manipulation?" In: *Behavioural Ecology: An Evolutionary Approach*, ed. J. R. Krebs and N. B. Davies, pp. 282–309. Oxford: Blackwell Scientific, 1978.

3. Cheney, D. L., and R. M. Seyfarth. *How Monkeys See the World*. Chicago: University of Chicago Press, 1990.

4. Mitani, J. C., et al. "Dialects in wild chimpanzees?" *American Journal of Primatology* 27 (1992): 233–243.

5. Silk, J. B., E. Kaldor, and R. Boyd. "Cheap talk when interests conflict." *Animal Behaviour* 59 (2000): 423–432.

6. Green, S. "Communication by a graded vocal system in Japanese monkeys." In: *Primate Behavior,* ed. L. A. Rosenblum, pp. 1–102. New York: Academic Press, 1975.

7. Fichtel, C., S. Perry, and J. Gros-Louis. "Alarm calls of white-faced capuchin monkeys: An acoustic analysis." *Animal Behaviour* 70 (2005): 165–176.

8. Gros-Louis, J. "The function of food-associated calls in white-faced capuchin monkeys, *Cebus capucinus,* from the perspective of the signaller." *Animal Behaviour* 67 (2004): 431–440.

9. Gros-Louis, J. "Responses of white-faced capuchins (*Cebus capucinus*) to naturalistic and experimentally presented food-associated calls." *Journal of Comparative Psychology* 118 (2004): 396–402.

10. Boinski, S., and A. F. Campbell. "The huh vocalization of white-faced capuchins: A spacing call disguised as a food call?" *Ethology* 102 (1996): 826–840.

11. Di Bitetti, M. S. "Food-associated calls and audience effects in tufted capuchin monkeys, *Cebus apella nigritus.*" *Animal Behaviour* 69 (2005): 911–919.

12. Boinski, S., and A. F. Campbell. "Use of trill vocalizations to coordinate troop movement among white-faced capuchins: A second field test." *Behaviour* 132 (1995): 875–901.

13. Gros-Louis, J. "Contexts and behavioral correlates of trill vocalizations in wild white-faced capuchin monkeys." *American Journal of Primatology* 57 (2002): 189–202.

14. Owren, M. J., et al. "Vocalizations of rhesus (*Macaca mulatta*) and Japanese (*M. fuscata*) macaques cross-fostered between species show evidence of only limited modification." *Developmental Psychobiology* 26 (1993): 389–406.

15. Hauser, M. D., N. Chomsky, and W. Tecumseh Fitch. "The faculty of language: What is it, who has it, and how did it evolve?" *Science* 298 (2002): 1569–1579.

16. Cleveland, J., and C. T. Snowdon. "The complex vocal repertoire of the adult cotton-top tamarin (*Saguinus oedipus oedipus*)." *Zeitschrift fuer Tierpsychologie* 58 (1982): 231–270.

17. Robinson, J. G. "Vocal systems regulating within-group spacing." In: *Primate Communication,* ed. C. T. Snowdon, C. H. Brown, and M. R. Petersen, pp. 94–116. New York: Cambridge University Press, 1982.

18. Zuberbühler, K. "A syntactic rule in forest monkey communication." *Animal Behaviour* 63 (2002): 293–299.

19. Arnold, K., and K. Zuberbühler. "Semantic combinations in primate calls: Putty-nosed monkeys rely on two basic calling sounds to construct a message of utmost urgency." *Nature* 441 (2006): 303.

20. Rowell, T. E. "Hierarchy in the organization of a captive baboon group." *Animal Behaviour* 14 (1966): 430–443.

21. De Waal, F., and A. van Roosmalen. "Reconciliation and consolation among chimpanzees." *Behavioral Ecology and Sociobiology* 5 (1979): 55–66.

22. Aureli, F., M. Cords, and C. P. Van Schaik. "Conflict resolution following aggression in gregarious animals: A predictive framework." *Animal Behaviour* 64 (2002): 325–343.

23. De Waal, F. B. M. "Conflict as negotiation." In: *Great Ape Societies*, ed. W. C. McGrew et al., pp. 159–172. Cambridge: Cambridge University Press, 1996.

24. Cords, M., and F. Aureli. "Reconciliation and relationship qualities." In: *Natural Conflict Resolution*, ed. F. Aureli and F. B. M. de Waal, pp. 177–198. Berkeley: University of California Press, 2000.

25. Cords, M., and S. Thurnheer. "Reconciling with valuable partners by long-tailed macaques." *Ethology* 93 (1993): 315–325.

26. Koyama, N. F. "The long-term effects of reconciliation in Japanese macaques (*Macaca fuscata*)." *Ethology* 107 (2001): 975–987.

27. Manson, J. H., S. Perry, and D. Stahl. "Reconciliation in wild white-faced capuchins (*Cebus capucinus*)." *American Journal of Primatology* 65 (2005): 205–219.

28. Oppenheimer, J. R. "Social and communicatory behavior in the *Cebus* monkey." In: *Behavioral Regulators of Behavior in Primates*, ed. C. R. Carpenter, pp. 251–271. Cranbury, NJ: Associated University Presses, 1973.

29. Perry, S. "Female-female social relationships in wild white-faced capuchin monkeys, *Cebus capucinus*." *American Journal of Primatology* 40 (1996): 167–182.

30. Carosi, M., et al. "Virilized external genitalia in female tufted capuchins (*Cebus apella*)." *American Journal of Primatology* 51 (2000): 50.

31. McDonald, D. B. "Cooperation under sexual selection: Age-graded changes in a lekking bird." *American Naturalist* 134 (1989): 709–730.

32. Colmenares, F. "Greeting behaviour in male baboons, I: communication, reciprocity and symmetry." *Behaviour* 113 (1990): 81–116.

33. Smuts, B. B., and J. M. Watanabe. "Social relationships and ritualized greetings in adult male baboons (*Papio cynocephalus anubis*)." *International Journal of Primatology* 11 (1990): 147–172.

34. Manson, J. H., S. Perry, and A. R. Parish. "Nonconceptive sexual behavior in bonobos and capuchins." *International Journal of Primatology* 18 (1997): 767–786.

35. Campos, F., J. H. Manson, and S. Perry. "Urine washing and sniffing in wild white-faced capuchins (*Cebus capucinus*): Testing functional hypotheses." *International Journal of Primatology* 28 (2007): 55–72.

36. Goffman, E. *Interaction Ritual: Essays on Face-to-Face Behavior*. New York: Doubleday Anchor, 1967.

37. Collins, R. "On the microfoundations of macrosociology." *American Journal of Sociology* 86 (1981): 984–1014.


### 6. Abby and Tattle: Two Females' Political Careers

1. Manson, J. H., and S. Perry. "Correlates of self-directed behavior in wild white-faced capuchins." *Ethology* 106 (2000): 301–317.
2. Wrangham, R. W. "An ecological model of female-bonded primate groups." *Behaviour* 75 (1980): 262–300.
3. Sterck, E. H. M., D. P. Watts, and C. P. van Schaik. "The evolution of female social relationships in nonhuman primates." *Behavioral Ecology & Sociobiology* 41 (1997): 291–309.
4. Perry, S. "Female-female social relationships in wild white-faced capuchin monkeys, *Cebus capucinus*." *American Journal of Primatology* 40 (1996): 167–182.
5. Manson, J. H., et al. "Dynamics of female-female relationships in wild *Cebus capucinus*: Data from two Costa Rican sites." *International Journal of Primatology* 20 (1999): 679–706.
6. Manson, J. H., et al. "Time-matched grooming in female primates? New analyses from two species." *Animal Behaviour* 67 (2004): 493–500.
7. Seyfarth, R. M. "A model of social grooming among adult female monkeys." *Journal of Theoretical Biology* 65 (1977): 671–698.
8. Schino, G. "Grooming, competition and social rank among female primates: A meta-analysis." *Animal Behaviour* 62 (2001): 265–271.
9. Kawai, M. "On the rank system in a natural group of Japanese monkeys. I, II." *Primates* 1 (1958): 111–148.
10. Sade, D. S. "Determinants of dominance in a group of free-ranging rhesus monkeys." In: *Social Communication among Primates*, ed. S. A. Altmann, pp. 99–114. Chicago: University of Chicago Press, 1967.
11. Hill, D. A. "The effects of demographic variation on kinship structure and behavior." In: *Kinship and Behavior in Primates*, ed. B. Chapais and C. M. Berman, pp. 132–150. Oxford: Oxford University Press, 2004.


### 7. Curmudgeon: The Career of an Alpha Male

1. Perry, S. "Male-female social relationships in wild white-faced capuchins (*Cebus capucinus*)." *Behaviour* 134 (1997): 477–510.
2. Perry, S. "Male-male social relationships in wild white-faced capuchins, *Cebus capucinus*." *Behaviour* 135 (1998): 139–172.
3. Perry, S. "A case report of a male rank reversal in a group of wild white-faced capuchins (*Cebus capucinus*)." *Primates* 39 (1998): 51–70.
4. Di Fiore, A., and P. Gagneux. "Molecular primatology." In: *Primates in Per-*

*spective*, ed. C. J. Campbell et al., pp. 369–393. Oxford: Oxford University Press, 2006.

5. Cowlishaw, G., and R. I. Dunbar. "Dominance rank and mating success in male primates." *Animal Behaviour* 41 (1991): 1045–1056.

6. Alberts, S., H. E. Watts, and J. Altmann. "Queuing and queue-jumping: Long-term patterns of reproductive skew in male savannah baboons, *Papio cynocephalus*." *Animal Behaviour* 65 (2003): 821–840.

7. Pope, T. R. "The reproductive consequences of male cooperation in the red howler monkey: Paternity exclusion in multi-male and single-male troops using genetic markers." *Behavioral Ecology & Sociobiology* 27 (1990): 439–446.

8. De Ruiter, J. R. "Male social rank and reproductive success in wild long-tailed macaques: Paternity exclusions by blood protein analysis and DNA-fingerprinting." In: *Paternity in Primates: Genetic Tests and Theories*, ed. R. D. Martin, A. F. Dixson, and E. J. Wickings, pp. 175–191. Basel: Karger, 1992.

9. Altmann, J., et al. "Behavior predicts genetic structure in a wild primate group." *Proceedings of the National Academy of Sciences of the United States of America* 93 (1996): 5797–5801.

10. Manson, J. H. "Mate choice." In: *Primates in Perspective*, ed. C. J. Campbell et al., pp. 447–463. Oxford: Oxford University Press, 2006.

11. Manson, J. H. "Measuring female mate choice in Cayo Santiago rhesus macaques." *Animal Behaviour* 44 (1992): 405–416.

12. Hamilton, W. D. "The genetical evolution of social behavior." *Journal of Theoretical Biology* 7 (1964): 1–51.

13. Soltis, J., et al., "Sexual selection in Japanese macaques, 2: Female mate choice and male-male competition." *Animal Behaviour* 54 (1997): 737–746.

14. Smuts, B. B. *Sex and Friendship in Baboons*, 1st Harvard paperback ed. Cambridge, MA: Harvard University Press, 1999.

15. Hrdy, S. B. *The Woman That Never Evolved*. Cambridge, MA: Harvard University Press, 1981.

16. Clutton-Brock, T. R., and P. H. Harvey. "Evolutionary rules and primate societies." In: *Growing Points in Ethology*, ed. P. P. G. Bateson and R. A. Hinde, pp. 195–237. Cambridge: Cambridge University Press, 1976.

17. Pusey, A., and C. Packer. "Dispersal and philopatry." In: *Primate Societies*, ed. B. B. Smuts, et al., pp. 250–266. Chicago: University of Chicago Press, 1987.

18. Clutton-Brock, T. H. "Female transfer and inbreeding avoidance in social mammals." *Nature* 337 (1989): 70–72.

19. Westermarck, E. *The History of Human Marriage*. London: Macmillan, 1891.

20. Shepher, J. "Mate selection among second-generation kibbutz adolescents and adults: Incest avoidance and negative imprinting." *Archives of Sexual Behavior* 1 (1971): 293–307.

21. Wolf, A. P. *Sexual Attraction and Childhood Association: A Chinese Brief for Edward Westermarck*. Stanford, CA: Stanford University Press, 1995.

22. Lieberman, D., J. Tooby, and L. Cosmides. "Does morality have a biological basis? An empirical test of the factors governing moral sentiments relating to incest." *Proceedings of the Royal Society Biological Sciences Series B* 270 (2003): 819–826.

23. Fessler, D. M. T., and C. D. Navarrete. "Third-party attitudes toward sibling incest: Evidence for Westermarck's hypotheses." *Evolution and Human Behavior* 25 (2004): 277–294.

24. Paul, A. "Sexual selection and mate choice." *International Journal of Primatology* 23 (2002): 877–904.

25. Manson, J. H., and S. E. Perry. "Inbreeding avoidance in rhesus macaques: Whose choice?" *American Journal of Physical Anthropology* 90 (1993): 335–344.

26. Soltis, J., et al. "Female mating strategy in an enclosed group of Japanese macaques." *American Journal of Primatology* 47 (1999): 263–278.

27. Chapais, B., and C. Mignault. "Homosexual incest avoidance among females in captive Japanese macaques." *American Journal of Primatology* 23 (1991): 171–183.

28. Smith, K., S. Alberts, and J. Altmann. "Wild female baboons bias their social behaviour towards paternal half-sisters." *Proceedings of the Royal Society Biological Sciences Series B* 270 (2003): 503–510.

29. Alberts, S. C. "Paternal kin discrimination in wild baboons." *Proceedings of the Royal Society of London B* 266 (1999): 1501–1506.

30. Fox, R. *The Red Lamp of Incest: A Study in the Origins of Mind and Society.* Chicago: Notre Dame University Press, 1983.

31. Watts, D. P. "Mountain gorilla life histories, reproductive competition, and socio-sexual behavior and some implications for captive husbandry." *Zoo Biology* 9 (1990): 185–200.

32. Smith, D. G. "Avoidance of close consanguineous inbreeding in captive groups of rhesus macaques." *American Journal of Primatology* 35 (1995): 31–40.

33. Kuester, J., A. Paul, and J. Arnemann. "Kinship, familiarity and mating avoidance in Barbary macaques, *Macaca sylvanus.*" *Animal Behaviour* 48 (1994): 1183–1194.

34. Muniz, L., et al. "Father-daughter inbreeding avoidance in a wild primate population." *Current Biology* 16 (2006): R156–AR157.

35. Janson, C. H. "Female choice and mating system of the brown capuchin monkey *Cebus apella* (Primates: Cebidae)." *Zeitschrift fuer Tierpsychologie* 65 (1984): 177–200.

## 8. Moth and Tranquilo: The Strategies of Incoming Alpha Males

1. Jack, K. M., and L. M. Fedigan. "Male dispersal patterns in white-faced capuchins, *Cebus capucinus*, 2: Patterns and causes of secondary dispersal." *Animal Behaviour* 67 (2004): 771–782.

2. Hrdy, S. B. "Male-male competition and infanticide among the langurs (*Presbytis entellus*) of Abu, Rajasthan." *Folia Primatologica* 22 (1974): 19–58.

3. Fedigan, L. M., "Impact of male takeovers on infant deaths, births, and conceptions in *Cebus capucinus* at Santa Rosa, Costa Rica." *International Journal of Primatology* 24 (2003): 723–741.

4. Van Schaik, C. P. "Infanticide by male primates: The sexual selection hypothesis revisited." In: *Infanticide by Males and Its Implications*, ed. C. P. van Schaik and C. H. Janson, pp. 27–60. Cambridge: Cambridge University Press, 2000.

5. Sommer, V. "The holy wars about infanticide: Which side are you on? And why?" In: *Infanticide by Males and Its Implications*, ed. C. P. van Schaik and C. H. Janson, pp. 9–26. Cambridge: Cambridge University Press, 2000.

6. Sussman, R. W., J. M. Cheverud, and T. Q. Bartlett. "Infant killing as an evolutionary strategy: Reality or myth?" *Evolutionary Anthropology* 3 (1995): 149–151.

7. Manson, J. H., J. Gros-Louis, and S. Perry. "Three apparent cases of infanticide by males in wild white-faced capuchins (*Cebus capucinus*)." *Folia Primatologica* 75 (2004): 104–106.


### 9. Kola and Jordan: Lethal Aggression and the Importance of Allies

1. Trivers, R. L. "Parental investment and sexual selection." In: *Sexual Selection and the Descent of Man*, ed. B. Campbell, pp. 136–179. Chicago: Aldine, 1972.

2. Emlen, S. T., and L. W. Oring. "Ecology, sexual selection, and the evolution of mating systems." *Science* 197 (1977): 215–223.

3. Wrangham, R. W. "An ecological model of female-bonded primate groups." *Behaviour* 75 (1980): 262–300.

4. Perry, S. "Intergroup encounters in wild white-faced capuchins (*Cebus capucinus*)." *International Journal of Primatology* 17 (1996): 309–330.

5. Jack, K. M. "Life history patterns of male white-faced capuchins (*Cebus capucinus*): Male bonding and the evolution of multimale groups." PhD diss., University of Alberta, 2001.

6. Jack, K. M., and L. M. Fedigan. "Male dispersal patterns in white-faced capuchins, *Cebus capucinus*. Part 1: Patterns and causes of natal emigration." *Animal Behaviour* 67 (2004): 761–769.

7. Jack, K. M., and L. M. Fedigan. "Male dispersal patterns in white-faced capuchins, *Cebus capucinus*. Part 2: Patterns and causes of secondary dispersal." *Animal Behaviour* 67 (2004): 771–782.

8. Gros-Louis, J., S. Perry, and J. H. Manson. "Violent coalitionary attacks and intraspecific killing in wild white-faced capuchin monkeys (*Cebus capucinus*)." *Primates* 44 (2003): 341–346.

9. Sussman, R. W., and P. Garber. "Cooperation and competition in primate so-

cial interactions." In: *Primates in Perspective*, ed. C. J. Campbell et al., pp. 636–651. Oxford: Oxford University Press, 2006.

10. Popp, J. L., and I. DeVore. "Aggressive competition and social dominance theory." In: *The Great Apes*, ed. D. A. Hamburg, and E. R. McCown, pp. 317–338. Menlo Park, CA: W. A. Benjamin, 1979.

11. De Waal, F. *Peacemaking among Primates*. Cambridge, MA: Harvard University Press, 1989.

12. Richerson, P. J., and R. Boyd. *Not by Genes Alone: How Culture Transformed Human Evolution*. Chicago: University of Chicago Press, 2004.

13. De Waal, F. B. M. *Our Inner Ape*. New York: Riverhead, 2005.

14. Wrangham, R. "The evolution of coalitionary killing." *Yearbook of Physical Anthropology* 42 (1999): 1–30.

15. Manson, J. H., and R. W. Wrangham. "Intergroup aggression in chimpanzees and humans." *Current Anthropology* 32 (1991): 369–390.

16. Wilson, M. L., M. D. Hauser, and R. Wrangham. "Does participation in intergroup conflict depend on numerical assessment, range location, or rank for wild chimpanzees?" *Animal Behaviour* 61 (2001): 1203–1216.

17. Aureli, F., et al. "Raiding parties of male spider monkeys: Insights into human warfare?" *American Journal of Physical Anthropology* 131 (2006): 486–497.

18. Tooby, J., and L. Cosmides. "The evolution of war and its cognitive foundations." Institute for Evolutionary Studies Technical Report 88-1.

19. Daly, M., and M. Wilson. *Homicide*. New York: Aldine de Gruyter, 1988.

20. Wrangham, R., and D. Peterson. *Demonic Males: Apes and the Origins of Human Violence*. Boston: Houghton Mifflin, 1996.

21. Sherif, M., et al. *Intergroup Conflict and Cooperation: The Robbers Cave Experiment*. Norman: University of Oklahoma Book Exchange, 1961.

22. Kurzban, R., J. Tooby, and L. Cosmides. "Can race be erased? Coalitional computation and social categorization." *Proceedings of the National Academy of Sciences* 98 (2001): 15387–15392.

### 10. Miffin, Nobu, and Abby: Capuchin Mothers, Infants, and Babysitters

1. Hartwig, W. C. "Perinatal life history traits in New World monkeys." *American Journal of Primatology* 40 (1996): 99–130.

2. Maestripieri, D. "Social structure, infant handling, and mothering styles in group-living Old World monkeys." *International Journal of Primatology* 15 (1994): 531–554.

3. Symons, D. "Beauty is in the adaptations of the beholder: The evolutionary psychology of human female sexual attractiveness." In: *Sexual Nature/Sexual Culture*, ed. P. R. Abramson and S. D. Pinkerton, pp. 80–120. Chicago: University of Chicago Press, 1995.

4. Digby, L. J. "Infant care, infanticide, and female reproductive strategies in polygynous groups of common marmosets (*Callithrix jacchus*)." *Behavioral Ecology and Sociobiology* 37 (1995): 51–61.

5. Silk, J. B. "Why are infants so attractive to others? The form and function of infant handling in bonnet macaques." *Animal Behaviour* 57 (1999): 1021–1032.

6. Lancaster, J. "Play-mothering: The relations between juvenile females and young infants among free-ranging vervet monkeys (*Cercopithecus aethiops*)." *Folia Primatologica* 15 (1971): 161–182.

7. Riedman, M. L. "The evolution of alloparental care and adoption in mammals and birds." *Quarterly Review of Biology* 57 (1982): 405–435.

8. Hrdy, S. B. "Care and exploitation of nonhuman primate infants by conspecifics other than the mother." In: *Advances in the Study of Behavior*, ed. J. S. Rosenblatt et al., pp. 101–158. New York: Academic Press, 1976.

9. Silk, J. B. "Kidnapping and female competition among captive bonnet macaques." *Primates* 21 (1980): 100–110.

10. Wasser, S. K. "Reproductive competition and cooperation among female yellow baboons." In: *Social Behavior of Female Vertebrates*, ed. S. K. Wasser, pp. 349–390. New York: Academic Press, 1983.

11. Manson, J. H. "Infant handling in wild *Cebus capucinus*: Testing bonds between females?" *Animal Behaviour* 57 (1999): 911–921.

12. Perry, S. "Female-female social relationships in wild white-faced capuchin monkeys, *Cebus capucinus*." *American Journal of Primatology* 40 (1996): 167–182.

13. O'Brien, T. G. "Parasitic nursing behavior in the wedge-capped capuchin monkey (*Cebus olivaceus*)." *American Journal of Primatology* 16 (1988): 341–344.

14. Perry, S., and J. C. Ordoñez J. "The effects of food size, rarity, and processing complexity on white-faced capuchins' visual attention to foraging conspecifics." In: *Feeding Ecology in Apes and Other Primates*, ed. G. Hohmann, M. M. Robbins, and C. Boesch, pp. 203–234. Cambridge: Cambridge University Press, 2006.

15. Henzi, S. P., and L. Barrett. "Infants as a commodity in a baboon market." *Animal Behaviour* 63 (2002): 915–921.

16. Zahavi, A. "The testing of a bond." *Animal Behaviour* 25 (1977): 246–247.

17. Zahavi, A. "Mate selection: A selection for a handicap." *Journal of Theoretical Biology* (1975): 205–214.

18. Grafen, A. "Biological signals as handicaps." *Journal of Theoretical Biology* 144 (1990): 517–546.

19. Smuts, B. B., and J. M. Watanabe. "Social relationships and ritualized greetings in adult male baboons (*Papio cynocephalus anubis*)." *International Journal of Primatology* 11 (1990): 147–172.

20. Gros-Louis, J., et al. "Vocal repertoire of white-faced capuchin monkeys (*Cebus capucinus*): Acoustic structure, context and usage" (forthcoming).

21. Chapais, B. "Alliances as a means of competition in primates: Evolutionary, developmental, and cognitive aspects." *Yearbook of Physical Anthropology* 38 (1995): 115–136.

22. Fairbanks, L. A. "Communication of food quality in captive *Macaca nemestrina* and free-ranging *Ateles geoffroyi*." *Primates* 16 (1975): 181–190.

23. Dewar, G. "Innovation and social transmission in animals: A cost-benefit model of the predictive function of social and nonsocial cues." PhD diss., University of Michigan, 2003.

## 11. Guapo: Innovation and Tradition in the Creation of Bond-Testing Rituals

1. McClintock, M. K. "Estrous synchrony and its mediation by airborne chemical communication (*Rattus norvegicus*)." *Hormones and Behavior* 10 (1978): 264–276.

2. McClintock, M. K. "Menstrual synchrony and suppression." *Nature* 229 (1971): 244–245.

3. Nicholson, B. "Does kissing aid human bonding by semiochemical addiction?" *British Journal of Dermatology* 111 (1984): 623–627.

4. Fragaszy, D. M., and S. Perry, eds. *The Biology of Traditions: Models and Evidence*. Cambridge: Cambridge University Press, 2003.

5. Panger, M., et al. "Cross-site differences in the foraging behavior of white-faced capuchin monkeys (*Cebus capucinus*)." *American Journal of Physical Anthropology* 119 (2002): 52–66.

6. Rose, L. M., et al. "Interspecific interactions between white-faced capuchins (*Cebus capucinus*) and other species: Preliminary data from three Costa Rican sites." *International Journal of Primatology* 24 (2003): 759–796.

7. Perry, S., et al. "Social conventions in wild white-faced capuchin monkeys: Evidence for traditions in a neotropical primate." *Current Anthropology* 44 (2003): 241–268.

8. Zahavi, A. "The testing of a bond." *Animal Behaviour* 25 (1977): 246–247.

9. Smuts, B. B., and J. M. Watanabe. "Social relationships and ritualized greetings in adult male baboons (*Papio cynocephalus anubis*)." *International Journal of Primatology* 11 (1990): 147–172.

10. Collins, R. "On the microfoundations of macrosociology." *American Journal of Sociology* 86 (1981): 984–1014.

11. Collins, R. "Emotional energy as the common denominator of rational action." *Rationality and Society* 5 (1993): 203–230.

12. Perry, S., and J. H. Manson. "Traditions in monkeys." *Evolutionary Anthropology* 12 (2003): 71–81.

13. Whiten, A., et al. "Cultures in chimpanzees." *Nature* 399 (1999): 682–685.

14. Stephenson, G. R. "Testing for group-specific communication patterns in Japanese macaques." In: *Proceedings of the Fourth International Congress of Primatology*, pp. 51–75. Basel: Karger, 1973.

## 12. Social Learning and the Roots of Culture

1. Fragaszy, D. M., and S. Perry, eds. *The Biology of Traditions: Models and Evidence.* Cambridge: Cambridge University Press, 2003.

2. Box, H., and K. R. Gibson, eds. *Mammalian Social Learning: Comparative and Ecological Perspectives.* Cambridge: Cambridge University Press, 1999.

3. Heyes, C. M., and B. G. Galef, eds. *Social Learning in Animals: The Roots of Culture.* New York: Academic Press, 1996.

4. Whiten, A., et al. "Cultures in chimpanzees." *Nature* 399 (1999): 682–685.

5. Russon, A. E. "Developmental perspectives on great ape traditions." In: *The Biology of Traditions: Models and Evidence*, ed. D. M. Fragaszy and S. Perry, pp. 329–364. Cambridge: Cambridge University Press, 2003.

6. Panger, M., et al. "Cross-site differences in the foraging behavior of white-faced capuchin monkeys (*Cebus capucinus*)." *American Journal of Physical Anthropology* 119 (2002): 52–66.

7. Perry, S., and J. C. Ordoñez J. "The effects of food size, rarity, and processing complexity on white-faced capuchins' visual attention to foraging conspecifics." In: *Feeding Ecology in Apes and Other Primates*, ed. G. Hohmann, M. M. Robbins, and C. Boesch, pp. 203–234. Cambridge: Cambridge University Press, 2006.

8. Visalberghi, E. "Tool use in a South American monkey species: An overview of the characteristics and limits of tool use in *Cebus apella*." In: *The Use of Tools by Human and Non-Human Primates*, ed. A. Berthelet and J. Chavaillon. Oxford: Clarendon Press, 1993.

9. Fragaszy, D. M., E. Visalberghi, and L. M. Fedigan. *The Complete Capuchin: The Biology of the Genus Cebus.* Cambridge: Cambridge University Press, 2004.

10. Ottoni, E. B., B. D. Resende, and P. Izar. "Watching the best nut-crackers: What capuchin monkeys know about others' tool-using skills." *Animal Cognition* 8 (2005): 215–219.

11. Lonsdorf, E. V., L. E. Eberly, and A. Pusey. "Sex differences in learning in chimpanzees." *Nature* 428 (2004): 715–716.

12. Coussi-Korbel, S., and D. M. Fragaszy. "On the relation between social dynamics and social learning." *Animal Behaviour* 50 (1995): 1441–1453.

13. Van Schaik, C. P., R. O. Deaner, and M. Y. Merrill. "The conditions for tool use in primates: Implications for the evolution of material culture." *Journal of Human Evolution* 36 (1999): 719–741.

## 13. Nobu and *La Lucha sin Fin:* Conservation of Tropical Dry Forests

1. Salazar, R. "Environmental law of Costa Rica: Development and enforcement." In: *Biodiversity Conservation in Costa Rica: Learning the Lessons in a Seasonal Dry Forest*, ed. G. W. Frankie, A. Mata, and S. B. Vinson, pp. 281–288. Berkeley: University of California Press, 2004.

2. Bustos, J. A. "Dispute over the protection of the environment in Costa Rica." In: *Biodiversity Conservation in Costa Rica: Learning the Lessons in a Seasonal Dry Forest*, ed. G. W. Frankie, A. Mata, and S. B. Vinson, pp. 289–298. Berkeley: University of California Press, 2004.

3. Quesada, M., and K. E. Stoner. "Threats to the conservation of tropical dry forest in Costa Rica." In: *Biodiversity Conservation in Costa Rica: Learning the Lessons in a Seasonal Dry Forest*, ed. G. W. Frankie, A. Mata, and S. B. Vinson, pp. 266–280. Berkeley: University of California Press, 2004.

4. Stoner, K. E., and R. M. Timm. "Tropical dry-forest mammals of Palo Verde: Ecology and conservation in a changing landscape." In: *Biodiversity Conservation in Costa Rica: Learning the Lessons in a Seasonal Dry Forest*, ed. G. W. Frankie, A. Mata, and S. B. Vinson, pp. 48–66. Berkeley: University of California Press, 2004.

# Acknowledgments

The ongoing seventeen-year study that is the source for the events described in this book requires a large collaborative effort, and if we were to thank all of the people who deserve thanks, the acknowledgments would be as long as the book itself. So we will reluctantly restrain ourselves to thanking only those who have made truly spectacular investments in the project of a year or more. The three people on whom we have depended the most over the past three years are Julie Gros-Louis, Hannah Gilkenson, and Wiebke Lammers. Julie has been with us almost from the very beginning, and her aid was particularly crucial in maintaining the demographic database during our early years at the University of California, Los Angeles, when we could make only sporadic visits to the site. Hannah Gilkenson was our right-hand woman from 2001 to 2006, handling the logistical details of purchasing, the scheduling, and the supervision of field assistants, as well as collecting large amounts of data. We feel that she was the heart and soul of the monkey project during her time there. Wiebke Lammers was one of our most talented field assistants ever, and subsequently devoted a year and a half of her life to managing the database, conducting pilot analyses of the data, and coordinating the data reliability checks—a tedious but absolutely crucial job, which she performed with astounding efficiency and accuracy. She returned in 2006 to be Hannah's successor as field manager, and to collect data for her own thesis. We could not have maintained continuity over such long time periods, or managed such a large crew of assistants, without the expert assistance and dedication of these three women, all of whom brought extraordinary ethological talent and organizational skills to the project, in addition to an amazing aptitude for establishing cooperative relations with others.

Our three Costa Rican field assistants, Alex Fuentes Jiménez, Mino Fuentes Alvarado, and Juan Carlos (Juanca) Ordoñez Jiménez, each devoted several years to the project and have been particularly valuable not only in maintaining the long-term database but also in helping us establish productive relationships with the local community. The following additional individuals contributed a year of service to the project between 2002 and 2006: Lydia Beaudrot, Mackenzie Bergstrom, Anne Bjorkman, James Broesch, Jo Butler, Nando Campos, Cindy Carlson, Nic Donati, Gayle Dower, Colleen Gault, Irene Godoy, Susie Herbert, Laura Johnson, Marie Kay, Emily Kennedy, Daphné Kerhoas-Essens, Sharon Kessler, Tom Lord, Whitney Meno, Nick Parker, Brent Pav, Kristen Potter, Kevin Ratliff, Heidi Ruffler, Chris Schmitt, and Eva Wikberg. Cindy, Colleen, Marie, Susie, Juanca, and Daphné showed phenomenal dedication in going right back to work after their near-fatal car accident, which left some of them seriously injured.

Todd Bishop and Kathryn Atkins worked with us early on; they showed remarkable fortitude in cheerfully staying to the end of their six-month contracts in 1997, when we were camping out under very unpleasant conditions. Eva Wikberg and Susan Wofsy served as data analysts for the project for extended periods of time. Over the years, Norma Amaya Araya and her family, Minor Mendoza Fernandez, and the Rosales family have provided us with extensive logistical support as well as friendship and good humor.

We are grateful to the Costa Rican park service (in its various incarnations as SPN, MIRENEM, SINAC, MINAE, ACT, and ACAT), to the community of San Ramon de Bagaces (especially Daniel Rojas), to Hacienda Pelon de la Bajura (especially Don Antonio Loaíciga P.), and to Rancho Brin D'Amour, for giving us permission to study the monkeys and for protecting the forest in which the monkeys live.

Our project was funded entirely by the Max Planck Society from July 2001 to June 2006, and we were most grateful for this generous support, which allowed us to devote ourselves entirely to research and writing. During the earlier phases of the project, we were funded by the L. S. B. Leakey Foundation, the National Science Foundation (a graduate fellowship, a NATO postdoctoral fellowship [no. 9633991], and a POWRE grant [no. SBR-9870429]), the Wenner-Gren Foundation, the National Geographic Society, the UCLA Academic Senate, the Rackham (University of Michigan) Graduate School, the Evolution and Human Behavior Program, the University of Michigan Alumnae Society, the Killam Trust, and Sigma Xi. Any opinions, findings, and conclusions or recommendations expressed in this material are our own and do not necessarily reflect the views of our funding agencies. The color illustration with the caption "Abby's twins eagerly observe her foraging efforts" previously appeared in Susan Perry and Juan Carlos Ordoñez Jiménez, "The Effects of Food Size, Rarity, and Processing Complexity on White-faced Capuchins' Visual Attention to Foraging Conspecifics," in *Feeding Ecology in Apes and Other Primates*, edited by Gottfried Hohmann, Martha M. Robbins, and Christophe Boesch (New York: Cambridge University Press, 2006).

The staff of the Max Planck Institute (especially Eva Land, Claudia Nebel, Erwin Ruff, Birgit Schubert, and Franca Thiele) showed remarkable patience, creativity, and efficiency in helping us navigate the red tape necessary to accomplish our research in three languages and three countries. We extend an extra special thanks to Alex Burkhardt, our omniscient computer technician, who even flew to Costa Rica to solve our many complex technological difficulties.

Most of the first draft was written when we were in the process of moving from Germany and Costa Rica back to Los Angeles to resume our jobs at UCLA. Almost all of our earthly possessions were packed, and inaccessible, on a boat that was slowly crossing the Atlantic, and we were homeless and without day care for our daughter. Eve, Don, Emma, and Elisabeth Cohen saved the day by taking us in for the summer and giving us a peaceful and well-furnished place to write in their home. The Cohens' unsurpassable hospitality, child-entertainment assistance, sense of humor, and highly therapeutic evening chamber music sessions were of invaluable aid to us.

Barbara Smuts, Linda Perry, Julie Gros-Louis, and two anonymous reviewers read the book manuscript in its entirety and provided us with many helpful comments. We are particularly grateful to Barb Smuts for inspiring both of us to become primatologists in the first place, back in the 1980s, and for providing us with much intellectual guidance and moral support in the years since then. The following people read and commented on specific chapters: Gayle Dower, Doug Durian, Michael Fisher, Hannah Gilkenson, Susie Herbert, Wiebke Lammers, Andrea Liu, Nick Parker, Nancy Peters, Martha Robbins, Frans de Waal, and Eva Wikberg. We thank our editor, Michael Fisher, for his patience and advice throughout this project; our copyeditor, Julie Hagen, for her expert assistance in preparing the final manuscript; and Elizabeth Gilbert, for her help with the artwork. We would like to thank Filippo Aureli for suggesting the title of the book.

And, of course, we are grateful most of all for the monkeys themselves, for entertaining us, inspiring us, and providing us with all the material on which we have based our academic careers in recent years. We hope the stories in this book make it evident exactly how much we admire and appreciate these remarkable individuals.

# Index

Italic page numbers refer to black-and-white illustrations.